City Politics and Public Policy

EDITED BY

JAMES Q. WILSON

City Politics
and Public Policy

John Wiley & Sons, Inc.
New York · London · Sydney

Contributors

CHARLES R. ADRIAN
Professor of Political Science   University of California at Riverside

ROBERT L. CRAIN
*Senior Study Director, National Opinion Research Center
University of Chicago*

MARTHA DERTHICK
Assistant Professor of Government   Harvard University

HEINZ EULAU
Professor of Political Science   Stanford University

ROBERT EYESTONE
Assistant Professor of Political Science   University of Minnesota

EDMUND P. FOWLER
*Assistant Professor of Political Science
Glendon College of York University (Toronto)*

JOHN A. GARDINER
Assistant Professor of Political Science   University of Wisconsin

J. DAVID GREENSTONE
Assistant Professor of Political Science   University of Chicago

HERBERT JACOB
Professor of Political Science   University of Wisconsin

ROBERT L. LINEBERRY

Assistant Professor of Government   University of Texas

PAUL E. PETERSON

Assistant Professor of Political Science and Education
University of Chicago

DONALD B. ROSENTHAL

Assistant Professor of Political Science
State University of New York at Buffalo

JAMES J. VANECKO

Assistant Study Director, National Opinion Research Center
University of Chicago

OLIVER P. WILLIAMS

Professor of Political Science   University of Pennsylvania

JAMES Q. WILSON

Professor of Government   Harvard University

ROBERT C. WOOD

Under Secretary of Housing and Urban Development   Washington, D.C.

Contents

City Politics
and Public Policy

JAMES Q. WILSON

Introduction:
City Politics
and Public Policy

The study of American local government for some time has had a peculiarly procedural quality. Originally, students of local government were essentially students of municipal law or municipal administration; that is, they were concerned with the formal arrangements under which local affairs were conducted and especially with those aspects of local affairs that relate to the administration of services. Although, from time to time, journalists (and an occasional scholar) exposed bossism or corruption, serious writing was preoccupied with the way in which services were provided to citizens and especially with ways in which those services might be provided more efficiently.

After the Second World War, a radical shift in emphasis occurred. Attention moved from the administration of services to the management of conflict, from how government works to "who governs," and from the formal and legal arrangements to the informal and extralegal distribution of influence. Power rather than function became the central theoretical concept and much effort was devoted to explicating its meaning and discovering who wielded it and how. Above all, these studies sought to be realistic—to explain how things *really* got done, especially with respect

1

to major community issues such as public housing, urban renewal, race relations, and land usage. Not everyone agreed as to how these questions might best be answered. Some sought to discover a general and persistent allocation of influence in the community that arose out of control over various nonpolitical resources—wealth and status, primarily—and to prove that men who had a disproportionate share of these resources exercised a disproportionate influence over the course of public affairs. Since nonpolitical resources were largely under the control of nonpoliticians, it was inevitable that, to the extent such scholars were able to show anything at all, they showed businessmen and other economic notables being deferred to—at least rhetorically and sometimes in matters of substance. Other scholars examined concrete issues to assess the relative influence of public and nonpublic leaders and usually found that men holding public positions, either in government or in political parties, were more likely to make the crucial decisions and to make them because of their political interests, not because of their dependence on private notables.

At the same time, the image of the political machine and its bosses was refurbished; this served several purposes. For some, calling attention to the machine called attention to the crucial problem of power and the inability of reformers to seize it or find an adequate substitute for it. To others, the machine represented an imperfect but serviceable agency for representing certain kinds of interests—primarily, those of low-income and ethnically segregated voters—which "good government" forms neglected. And for still others, the existence of the machine, or the strong political party generally, showed that politics could operate independently of the distribution of wealth and status in the community; that politics had a certain autonomy and was not simply, in Marxist terms, the epiphenomenon of economic life.

Although the newer studies seemed—and to a great extent were—more "realistic," they were still primarily concerned with governmental procedures, the difference being that attention had shifted from the way administrators managed their affairs to the way party and civic leaders struggled for power. It had changed, that is to say, from the conduct of bureaucrats to the tactics of politicians. But this new realism, perhaps because it *was* new, or perhaps because it was involved with its own methodological uncertainties, rarely had much to say about the functions of government or how political forms affected the way those functions were performed. We began to learn a great deal about who governs but surprisingly little about what *difference* it makes who governs. The average citizen, we might suppose, was not much interested in knowing with great precision the shape of the deference hierarchy in his community and only a few citizens were deeply affected by how a particular issue was resolved or suppressed. Indeed, in the two areas where issues might well engage the general interest of the citizen—law enforcement and the public schools—there were scarcely any scholarly studies at all. Most issues studied involved winning office or controlling the use of land (and not much land at that).[1]

One reason for this neglect might be that those local affairs of greatest concern to the citizen were rarely handled in the form of issues, they were handled in the form of taxes and services. In rejecting the older ways of studying public finance and public services, American scholars may well have (unintentionally) rejected the subjects of those studies as well.

It is the purpose of this book to encourage research on local government that makes public policy a central concern.[2] I believe that such a concern has traditionally been part of political science since Aristotle attempted to show the capacity of various regimes to produce virtue among their citizens. The best empirical political science has, in my view, usually (not always) been that which has tried to explain why one goal rather than another is served by government, and the consequences of serving that goal, or serving it in a particular way. Such a concern draws together the empirical and philosophical aspects of political inquiry so that those who try to explain why something is as it is and those who speculate on whether it should be as it is might reasonably be regarded as members of the same discipline.

This volume is not the first to assert an interest in the policy outcomes of local government. The book, *Four Cities,* by Oliver P. Williams and Charles R. Adrian made in 1963 a systematic effort to develop comparative empirical data on the public policies of four Michigan cities. A selection from it is included in this collection. From time to time, various studies have appeared which attempt to find associations between various policies (such as urban renewal programs or fluoridated water supplies)

[1] Some speculation on the policy implications of various local political systems was offered in Edward C. Banfield and James Q. Wilson, *City Politics* (Cambridge: Harvard University Press, 1964), the final chapter. A summary of some empirical findings is offered in Lewis A. Froman, "An Analysis of Public Policies in Cities," *Journal of Politics,* XXIX (February, 1967), pp. 94–108.

[2] I have previously developed the argument that follows in an essay entitled "Problems in the Study of Urban Politics," in Edward Buehrig, ed., *Essays in Political Science* (Bloomington: University of Indiana Press, 1966).

and the socioeconomic characteristics of cities.[3] A similar and more extensive effort has been made (using states as the unit of analysis) to explain differences in the level of, say, welfare expenditures.[4]

This demographic approach has relied for the most part on readily available (perhaps *too* readily available) census materials concerning the composition of local populations to obtain factors which might be thought of as causes of policy differences. Although we might readily agree that rich cities are likely to act differently than poor ones, there are two difficulties with this emphasis. The first is that it directs attention away from local governmental arrangements, political history and culture, party activities, and the political attitudes of key participants. The relative importance of these factors, as against underlying socioeconomic factors, is rarely judged. Instead, a crosstabulation of demographic and policy factors shows "some" relationship (or "no" relationship); how much a relationship, or how much *more* a relationship might be shown by introducing governmental factors, is not indicated. Or in a regression equation, a certain (usually modest) amount of the city-by-city variation is "explained" by a set of (rarely independent) variables; the unexplained variation is treated as a residual or error factor.

Some studies have attempted to cope with this by looking for governmental differences that might account for the residuals; others have had considerable success in reducing the size of the unexplained variation by searching out additional demographic variables. But this only gives rise to the second difficulty: in what sense have we "explained" a public policy by observing its association with certain population characteristics? That such characteristics are relevant is beyond much doubt, but how or why are they relevant? If what is to be explained is a set of individual behaviors—voting, for example—then a strong statistical association be-

---

[3] See Amos Hawley, "Community Power and Urban Renewal Success," *American Journal of Sociology*, January, 1963, pp. 422–431, and Maurice Pinard, "Structural Attachments and Political Support in Urban Politics: The Case of Fluoridation," *American Journal of Sociology*, March, 1963, pp. 513–526. A rejoinder to Hawley is Bruce C. Straits, "Community Adoption and Implementation of Urban Renewal," *American Journal of Sociology*, July, 1965, pp. 77–82.
[4] See for example Richard E. Dawson and James A. Robinson, "The Politics of Welfare," Chap. 10 in Herbert Jacob and Kenneth N. Vines, eds., *Politics in the American States* (Boston: Little Brown & Co., 1965) and Richard I. Hofferbert, "The Relation Between Public Policy and Some Structural and Environmental Variables in the American States," *American Political Science Review*, LX (March, 1966), pp. 73–82. Some of these studies have used certain political variables, such as party competition, apportionment, and the like, but few have used any measures of the distribution of influence.

tween behavior and characteristics is very significant. We must still assume—and ideally explore—the psychological state that leads a group of upper-income white Protestant businessmen to vote Republican, but the assumptions are not hard to imagine and the exploration is not, in principle, very difficult. But if what is to be explained is a collective decision, or a state of governmental affairs that may have existed for many years, or a developmental sequence with respect to the way a certain governmental function is managed, the assumptions linking even highly-significant demographic correlations with the observed outcomes are harder both to imagine and to analyze. Unless we are willing to assume that a given distribution of preferences among voters that arises out of the socioeconomic characteristics of those voters is directly and faithfully translated into public policy, then we must explain why politicians and bureaucrats choose some preferences rather than others, or impute to certain groups one preference rather than another, or even ignore public preferences in favor of what leaders think is good for the community.

One answer to this line of reasoning is that if we can accurately predict policies from demography (or from anything else, for that matter), then we have accomplished our task as social scientists and, unless someone devises a model that predicts better or as well but with less data and fewer assumptions, we have exhausted the meaning of the word "explanation," scientifically understood. There is great force in this argument and for many purposes it is quite satisfactory. Unhappily (or happily) we almost never do achieve such elegance—except occasionally and for aggregates of individual behaviors—and thus the issue does not arise. But even should it arise, I suspect that human curiosity is not so easily satisfied and most of us would still want to understand the political linkages between demography (or attitudes) and policy, especially since many of us would suspect that in a changing world a model highly predictive today may be hopelessly unpredictive ten years from now because "something has happened."

Thus, without denying the importance of continued efforts to find, statistically, the socioeconomic boundaries within which public policy varies from place to place, the studies in this book have been selected because they try to explain directly the impact of city government on urban policies. The single, dramatic case study has been avoided—no paper here presented involves fewer than two cities and some involve several score. The concern is for what is *generally* true, not what may have been the case in one place at one time. Those authors with either scarce resources or a desire to see for themselves have limited themselves to two, three, or four places and thus have been forced to suggest what may be generally true by describing the extreme cases between which the typical can be found or by indicating the ways in which some "typical" cities may differ. Other authors, with more generous resources or a research strategy that permits the use of more routinized data-gathering procedures, have examined a large number of cities to test systematically the effect of particular variables in accounting for the differences in a representative sample.

Furthermore, most of these studies are about routine government services, or issues concerning those services, and not about major community conflicts. (There are three exceptions—the papers on school desegregation, fluoridation, and the war on poverty.) The emphasis is on how cities differ in tax monies raised, planning programs funded, traffic tickets issued, welfare programs administered, juvenile misconduct controlled, and personal bankruptcies handled. The reader who is looking for excitement, human interest, and urban drama (or urban comedy) had better stop right here. We find the effort to explain what is generally true about, or what seems to account for differences among, various common government programs quite fascinating, but we are under no illusion that this is what would be written by a journalist reporting the latest dispatches from the urban battlefront. A journalist is properly interested when man bites dog. The authors of these papers are interested in the normal case, which is that dogs bite men, and in how the depth of the bite varies with the size of the dog, the thickness of the pants' leg, and the agility of the man.

The principal methodological problem here is the familiar one of index formation—that is, of finding some measure (verbal or quantitative) that is an accurate description for each member of a class of phenomena and thus can be used as a description of the class itself. Voting studies were the first major application of quantitative techniques to political science, in great part because the index problem was so easily solved—a vote cast is an unambiguous, homogeneous (that is, comparable from person to person), additive statement of particular behavior. Groups of voters can be compared by the votes they cast; individual voters by the votes they *said* they cast. But what index do we use to measure the output of a group of traffic policemen, welfare agencies, community action programs, school boards, or courts? Some of the most "obvious" measures turn out to be the trickiest. We might expect that taxes collected or expenditures made are nice, "hard" measures of something, but on closer examination it turns out that what a city collects or spends depends very much on state grants-in-aid, user charges, community resources, and the like, and thus though dollar amounts may be good measures it is not at all clear what they measure. In these studies, this problem has been handled in various ways. Some authors have looked for a long time at a few cases and given a summary description for each case on the ground that "this is the way it looks to me and would probably look to anybody else in my position." Others have found some numbers (for example, traffic tickets per thousand population) which, though individually suspect, can be used as rough measures of at least *gross* differences between cities. Still other authors have fashioned their own measures by (for example) surveying attitudes through questionnaires that inevitably miss much of the richness or subtlety of the respondents' states of mind. The problem of index formation often leads to what might be called the Index Dilemma: the more accurate the statement of some behavior, the fewer such behaviors can be observed, and thus the less the possibility of rigorously testing alternative explanations for such behavior.

Substantively, no sweeping generalizations can be offered on the basis of these studies, or if they can they are not apparent to the editor. But certain major issues arise in many of the studies and to these we call attention. The first and most obvious is the extent to which local government has an independent effect on public policy in ways that can be associated with the characteristics of that government, the party arrangements which animate it, and the values which permeate it. On some matters the answer is quite clear. Rosenthal and Crain show that, among large American cities, those with partisan governments are much more likely to fluoridate their water supplies than those with nonpartisan governments. It is not, the authors speculate, so much the partisan label on the ballot as the concentration of political power brought about by strong parties that accounts for the ability of these cities to adopt fluoridation despite the controversy that surrounds it. Among middle-sized cities, the same process seems to be at work, though there political power can be concentrated in ways other than by strong parties. A city with a strong, professionally oriented city manager is more likely than one with a weak, nonpartisan mayor to adopt fluoridation; so is a city with a strong, partisan mayor, especially one whose affiliation is with a national political party rather than simply with a local one.

Similarly, Greenstone and Peterson show that, among the four largest cities at least, political power concentrated in the hands of a strong mayor enables the city to respond quickly to the opportunity to get federal anti-poverty money and to get a lot of that money but lessens the opportunity for the poor to achieve "maximum feasible participation" in the local community action program. Crain and Vanecko indicate that in eight large cities, school desegregation decisions are made by those who are supposed to make them—school boards—and not by economic or political forces operating behind the scenes, and that school boards that are appointive rather than elective are more likely to acquiesce in desegregation plans. An appointive board, far from being "political" in the sense meant by reformers who fear the domination of a mayor, is nonpolitical in the sense that it is *less* responsive to popular opinion, whether of defenders of neighborhood schools or militant advocates of civil rights, and more responsive to its own personal convictions in the matter. And these convictions, though perhaps formed by the board's prior associations, are not dictated by it. Paradoxically, the prospect of having a board willing to accept desegregation is enhanced by having a *strong* political party because such an organization reduces citizen participation in government and places the school board in a position where civic organizations and non-political elites may influence nominations and defend a "zone of indifference" within which the board may act as it chooses. The difficulty with asserting too confidently such a generalization is that it may all be a matter of degree—a *very* strong party may dominate the school board to the exclusion of all civic influences, while a very weak party may allow the board to become a cockpit for unaffiliated personalities. What is required is a party that is strong enough and also inclined to see its self-interest in making the school board sufficiently autonomous so that the party need not take the blame for controversies in this sensitive area.

But perhaps these formal and informal governmental arrangements are themselves but the expression of underlying community characteristics. To some extent they are, and a few studies (none included here) have suggested that, for example, a city with a young, geographically mobile population is more likely to have a council-manager form of government than an older, more stable city.[5] Though this and other relationships exist in the data, they are, with a few exceptions, not very compelling—the

[5] Leo F. Schnore and Robert R. Alford, "Forms of Government and Socio-economic Characteristics of Suburbs," *Administrative Science Quarterly,* June, 1963, pp. 1–17; Robert Alford and Harry Scoble, "Political and Socioeconomic Characteristics of American Cities," *The Municipal Yearbook, 1965,* pp. 82–97; John H. Kessel, "Governmental Structure and Political Environment," *American Political Science Review,* LVI (September, 1962), pp. 615–620.

statistical correlations, though perhaps significant, are often weak. Lineberry and Fowler find no important demographic differences between "reformed" and "unreformed" cities—between, that is to say, cities with nonpartisan elections and city managers on the one hand and those with partisan elections and ward-based councils on the other. But reformed and unreformed cities *behave* differently on tax and expenditure matters—in the latter, variations in monies raised and spent seem strongly associated with underlying community characteristics, such as the socioeconomic composition of the population, while in the former such differences are much less strongly associated with community characteristics. A local government, apparently, can either let community preferences determine such matters (presumably by the pulling and hauling among groups or the regular testing of voter sentiment) or can act aggressively on its own without reference to (or in spite of) presumed preferences. A "reformed" government, perhaps, tends to substitute bureaucratic for political premises in the making of public finance decisions.

This conception of local politics as an intervening variable between community characteristics and public policy is further strengthened by Eyestone and Eulau who study city council members and decisions in a large number of California cities, all of which have more or less similar institutional forms. The political variables at which they look are the values and attitudes of the councilmen and the nature of the civic life of the community. They find that these factors are independent of certain characteristics of the community but have an influence on certain policy outcomes. More specifically, they suggest that how much money a city can "afford" to spend on planning and amenities (as judged by its assessed property valuation) does not determine what the councilmen believe it *should* spend, but what the councilmen believe—more precisely, whether or not they are committed to the idea of community development—does have an independent effect on what the community *will* spend.

All this may strike a reader who has had the good fortune not to profess political science for a living as all too obvious. If it is obvious, I think it is still important, but frankly I am not sure it is all that obvious. We might imagine that different kinds of voters prefer different kinds of governing institutions; that a majority gets the kind it wants; and that these institutions make sure the voters get the policies they want. In fact, what kind of governmental institutions a city has may be imposed by a state constitution or have been decided in a forgotten historical period. And what policies the government supplies may depend crucially on how power is distributed within it, who gets access to it, and what concrete choices are available in a particular case.

One of the most intriguing aspects of this situation is the relationship between political decision and administrative choice. This is the second main theme to emerge from the studies. Political scientists have for so long been denying the existence of any radical distinction between policy and administration that perhaps we have inadvertently come to suppose that there is no administrative process which is not somehow involved, and rather crucially involved, in the political process. We are right if we mean that few, if any, administrators make decisions that require no major value choices; we are wrong if we mean that every administrator chooses that value from his relationship or sensitivity to the surrounding political struggle. Gardiner suggests that police chiefs alone decide, within very broad limits, what traffic law enforcement policy a city will have. Neither community characteristics nor the intervention of political officials seems to have much general effect. Derthick notes that despite great differences in community characteristics and local political arrangements, the locally appointed welfare directors in Massachusetts cities and towns exercise relatively little discretion (they follow pretty closely state and federal norms) but such discretion as they do exercise seems to be in accordance with their own preferences in the matter.

In these two fields, at least, local administrators enjoy considerable freedom from local political demands, and certainly not because the average voter has no views on law enforcement or welfare programs. Nor is the autonomy the result of the high degree of professionalism of the administrators; police chiefs and welfare directors are scarcely the equivalent of the government physicist who does what he wants because it is "right" and his colleagues approve, and public opinion be damned. The autonomy police chiefs and welfare directors enjoy seems to be the result of the absence of any political incentives for elective officials to set *general* policies in those areas (though there may be powerful incentives for them to intervene on behalf of *particular* constituents) and because, in the case of traffic enforcement, the function has a very low visibility and, in the case of welfare, is tightly constrained by state and federal rules. Williams and Adrian imply something similar when they argue that the real distinction between policy and administration is a distinction between what is controversial and open to change and what is agreed-to or beyond deliberate or planned change. Being beyond change may mean not only being bound by rules set by others over whom the individual has no control but also being in the province of a large organization. Derthick indicates that above a certain threshold size, large bureaucracies display significantly more administrative discretion than small ones performing the same function. Bigness makes subordinates less visible and thus more difficult to control; big organizations are more likely to attract into key positions "professionals" who bring with them their own ideas as to how things should be done; and big organizations stimulate the formation of "watchdog" or client organizations on the outside whose pressures and interventions can cause deviations from official policy.

This may explain why Lineberry and Fowler find "reformed" city governments less sensitive to community characteristics (and presumably to immediate voter interests) with respect to budget policies. The reformed government has, perhaps, placed more power in the hands of administrators who on routine service matters have considerable autonomy and who become a source of policy innovation; city councils may find themselves voting for expenditures partly in response to these administrative pressures (and the pressures of client groups the administrators have helped bring into being) rather than simply in response to what they think the voters "want" or will tolerate.

Indeed, politicians may bring about the very pressures they later seek to avoid by the various strategies they employ to handle critical tax issues. Wood describes the ways in which communities in the 22-county New York metropolitan region grapple with the need for municipal revenue One strategy is, of course, to use land-use controls, notably zoning, to avoid the kind of growth in a community that can produce demands for services. But two other strategies, though they may have short-run benefits, are not so clearly in the long-run interests of the economizers. One is the creation of new governmental units—special districts or authorities—to provide services and tap new revenues; the other is to seek funds from higher levels of government. But both strategies call into being or bring about the intervention of other bureaucracies, and though these other organizations may pay their own way for a while, in the long run they are likely themselves to create pressures for additional local programs and services by co-opting local elites and stimulating the formation of clientele groups.

Another theme that several papers raise concerns the "political culture" of the community and its effect on public policy. Few terms in recent political science have become as fashionable as this, and few are as difficult to define precisely. This is not the place, and I am not the person, to clarify the concept. Suffice it to say that however maddeningly vague it may be, we seem increasingly unable to do without it. Very broadly, a political culture might be thought of as a widely shared, patterned view of the proper scope and behavior of public institutions and specifically of what ways of behaving on public matters (getting votes, casting votes, proposing programs, administering services, managing conflict) would be thought legitimate. We all have said, at one time or another, that a city or organization has a "climate" or "style" in public matters and that people—even those who may wish it were different—have come to expect things to be done that way. In one place a city bureaucrat may be brisk, in another folksy; some politicians compete to see who will keep taxes the lowest, others to see who will dream up the boldest new program. Williams and Adrian have given us a typology of four possible roles which government might play and find that they can fit into these categories descriptions of how at least a few cities do in fact behave. They suggest that underlying the dominant political culture of a community is a class-based conception of what government should do. That culture may change over time as people move in and out and as new issues arise. Crain and Vanecko suggest that a movement of the upper-class business elite to the suburbs has a profound effect on the political culture of the big central city, at least on school matters.

The critical problems in the use of the concept of political culture (other than defining it) are in finding a good measure of it and in showing a linkage between that culture and the behavior of governmental institutions. Such a linkage is rather obvious when what is being explained is voting behavior. Many studies have shown the independent impact of ethnicity and religion on voters' preferences for candidates; some studies have suggested that within any given city there are competing subcultures which affect public policy by patterning the vote of various groups on expenditure and other referenda.[6] But these are the easy cases; here, a culture (or subculture) is equivalent to the patterned response to a concrete choice of an aggregate of individuals sharing certain common characteristics.

How does culture affect government when decisions are not made by voting? Linebarry and Fowler show that a political culture may become institutionalized and thereby affect tax and expenditure levels. In a "re-

[6] James Q. Wilson and Edward C. Banfield, "Public-Regardingness as a Value Premise in Voting Behavior," *American Political Science Review,* LVIII (December, 1964), pp. 876–887.

formed" city—one in which they argue that the "middle class ethos" has been institutionalized, environmental factors (such as the characteristics of the population) affect revenue policies but to a significantly lesser degree than in "unreformed" cities. Or put another way, those institutions embodying a particular political culture, once adopted, make government a more important intervening variable between environment and outcome.

Another way culture can affect government is, as Jacob suggests, through the willingness or unwillingness of citizens to invoke certain public remedies for private disputes. Specifically, he finds that in certain Wisconsin cities, no different by any obvious demographic measure from some other cities in that state, citizens are much less inclined to take debtor-creditor issues into court by seeking wage garnishments or by appealing to bankruptcy proceedings. Indeed, his evidence indicates that this privatization of issues extends to a whole range of civil matters with the result that, in Green Bay at least, the courts have a lot less to do. If the cities are the same in population composition, why do they have different cultures? And if the cultures were at one time differentiated, how do these differences persist—how, in effect, are they "learned"? Since Jacob's study draws in part on a larger study of the political culture of the same four cities, we shall in time learn a good deal more about these matters.

The final issue I find in these papers is one of policy. American local government clearly faces the same dilemma of democratic government generally—that of coping with the trade-off between power and responsiveness, between a capacity for innovation and a sensitivity to citizen interests. Centralized political systems are more likely to act and to act in ways many people would regard as enlightened with respect to fluoridating their water, integrating their schools, and getting money for the poor. Decentralized systems are more likely to let the voters decide by referendum on fluoridation (and this, in turn, means that fluoridation will probably be rejected), to allow various groups to have access to the school board but in ways that may prevent it from acting, or acting quickly, on a school desegregation plan, and to give the poor greater representation in a community action program but at the expense of material resources. Which system we prefer clearly has a great deal to do with what goals we cherish; the difficulty is that most of us probably want one system for certain matters and its opposite for others, and sadly enough we can't have it both ways.

But the issue is not even that simple. A great deal depends on the terms on which power is concentrated. Centralized in the hands of a strong, professional manager, it may produce a large and vigorous bureaucracy which seeks out and uses discretion to advance its conception of the public interest and to formulate new programs for consideration. Power centralized in the hands of a party boss or strong mayor may produce a subservient bureaucracy, one that rarely makes proposals for change on its own initiative but one which is quite open to citizen interventions and sensitive to the value of bending policy to shape neighborhood and group interests. The party leader may not do much but on those matters on which he does act, he can act decisively.

The eleven papers in the succeeding sections of this book are from two sources. Six of them were written for and delivered at the panel on state and local government (of which I was chairman) held at the 1966 annual meeting of the American Political Science Association; they are published here for the first time.[7] The other five are reprinted, with the permission of the authors and copyright holders, from various scholarly journals and books. They are rather arbitrarily arranged into substantive policy areas —mostly for the benefit of readers who are less interested in the theoretical considerations (which in fact are the dominant considerations of most of the authors) than in the policy issues themselves. The APSA authors, because they think it important, agreed to write under a common set of constraints—that the papers would report empirical findings on a comparative basis of the policy outcomes of various local political systems. I believe they will join with me in urging others to follow their example.

[7] They are the papers by Eyestone and Eulau, Crain and Vanecko, Gardiner, Derthick, Jacob, and Greenstone and Peterson.

PART I

Politics
and Community
Development

OLIVER P. WILLIAMS

CHARLES R. ADRIAN

Community Types
and Policy Differences

A TYPOLOGY OF LOCAL GOVERNMENT

Initially let us consider differing roles of local government. The concept "role of government" may be given at least two interpretations. It may refer to images of the *proper* role of government or to the *actual* role of government. A typology of roles based on the first interpretation will be outlined initially. Conversion of the typology for the second usage will be discussed subsequently.

The typology characterizes four different roles for local government: (1) promoting economic growth; (2) providing or securing life's amenities; (3) maintaining (only) traditional services; (4) arbitrating among conflicting interests. The development of the typology will proceed as follows: First, each role will be described, followed by a discussion of the relationship between the roles and forms of governmental structure; second, the convertibility of the typology for different uses will be suggested; finally, the operational character of the typology and criteria for its application will be discussed.

The first type characterizes government as *the promotion of economic growth*. The object of government is to see that the community grows

Reprinted from Oliver P. Williams and Charles R. Adrian, *Four Cities* (Philadelphia: University of Pennsylvania Press, 1963), pages 23–26, 272–288, with the permission of the copyright holder, the Trustees of the University of Pennsylvania.

in population and/or total wealth. Born of speculative hopes, nurtured by the recall of frontier competition for survival, augmented by the American pride in bigness, the idea that the good "thriving" community is one that continues to grow has been and still remains a widely held assumption in urban political thinking. It is essentially an economic conception, drawing an analogy between municipal and business corporations. Just as the firm must grow to prosper, so must the city increase in population, industry, and total wealth. But the parallel is not simply an analogy, for according to this view the ultimate vocation of government is to serve the producer.

Although this image is endorsed most vigorously by those specific economic interests which have a stake in growth, its appeal is much broader. The drama of a growing city infects the inhabitants with a certain pride and gives them a feeling of being a part of progress: an essential ingredient of our national aspirations. The flocking of people to "our" city is something like the coming of the immigrants to our national shores. It is a tangible demonstration of the superiority of our way of life in that others have voluntarily chosen to join us. Growth also symbolizes opportunity, not only economically but also socially and culturally. Large cities have more to offer in this respect than smaller ones.

Certain groups have such concrete stakes in growth that they are likely to be the active promoters of this image, while others in the city may give only tacit consent. The merchant, the supplier, the banker, the editor, and the city bureaucrats see each new citizen as a potential customer, taxpayer, or contributor to the enlargement of his enterprise, and they form the first rank of the civic boosters.

The specific policy implications of this image of the role of government are varied. Historically, local government was deeply embroiled in the politics of railroad location, land development, and utility expansion. But today city government is more handicapped than it was in the nineteenth century as an active recruiter of industry, for the vagaries of industrial location are determined more by economic market considerations than by the character of locally provided services. Although a number of ways exist in which economic growth may be served, producer-oriented political activity often expresses itself negatively; that is, nothing should be done that might hinder the community's growth. The city should have "a good reputation." Politics should be conducted in a low key. The image of stability and regularity in city finances must be assured. Friendliness toward business in general should be the traditional attitude of city officials.

But to some, growth itself is not a desirable goal for a community. Indeed, it is the very thing to be avoided. Growth breeds complexities which, in turn, deny the possibilities of certain styles of life. For some the role of local government is *providing and securing life's amenities.* "Amenities" is used here to distinguish between policies designed to achieve the comforts and the necessities of life as opposed to only the latter. Both comfort and necessity are culturally defined attributes of living standards, but the distinction is useful. Most communities recognize the demand for amenities in some fashion, but only an occasional community makes this the dominant object of collective political action. Many cities have their noise- and smoke-abatement ordinances, but the idea of creating a quiet and peaceful environment for the home is hardly the central purpose of government.

The policies designed to provide amenities are expressed by accent on the home environment rather than on the working environment—the citizen as consumer rather than producer. The demands of the residential environment are safety, slowness of traffic, quiet, beauty, convenience, and restfulness. The rights of pedestrians and children take precedence over the claims of commerce. Growth, far from being attractive, is often objectionable. That growth which is permitted must be controlled and directed, both in terms of the type of people who will be admitted and the nature of physical changes. It is essentially a design for a community with a homogeneous population, for civic amenities are usually an expression of a common style of living. Furthermore, because amenities are costly, the population must have an above average income. All expenditures for welfare are unwanted diversions of resources. Those asking for public welfare assistance are seeking necessities which other citizens provide for themselves privately. Consequently, this image finds expression chiefly in the middle- to upper-income residential suburbs.

However, as we descend the scale from larger to smaller cities, middle-class consumers are not necessarily isolated in a self-imposed suburban exile and their demands may become channeled toward the core city itself. In other places, a fortunately located large industrial facility may allow a city of persons with modest means to enjoy costly amenities. The Ford Motor Company's tax base supplies Dearborn, Michigan, with sidewalk snow removal and a summer resort. There are villages with iron mines within their limits, located in the Upper Peninsula of Michigan, which have free bowling alleys in the city hall and other rarely found city services.

Amenities are also likely to become a dominant concern of local government in the traditional small town threatened by engulfment from a nearby major urban complex. Such a town will not be characterized by the homogeneity of the suburb, nor will its particular kind of amenities require costly expenditures, for the amenities being secured are simply those which are a function of small-town life. Because growth of the town was slow, the necessary capital investments for urban life were amortized over a very long time. A sudden population influx threatens both the styles of life and the cost of maintaining that style. However, all small-town civic policies can hardly be characterized as the desire to preserve amenities, for conditions in many small towns make a mockery of that term. This leads us to our third image of the role of local government: that of *maintaining traditional services* or *"caretaker government."*

Extreme conservative views toward the proper role of government can be more effectively realized at the local than at any other level. The expansion of governmental functions often can be successfully opposed here by the expedient of shunting problems to higher levels. Such evasions of difficult problems give the appearance of successful resistance to the extension of government and thus reinforce the plausibility of maintaining a caretaker local-governmental policy.[1]

"Freedom" and "self-reliance of the individual" are the values stressed by this view. Private decisions regarding the allocation of personal resources are lauded over governmental allocation through taxation. Tax increases are never justified except to maintain the traditional nature of the community. This substitution of individual for public decisions emphasizes a pluralistic conception of the "good." The analogy to *laissez faire* economics is unmistakable. The caretaker image is associated with a policy of opposition to zoning, planning, and other regulations of the use of real property. These sentiments are peculiarly common to the traditional small town.

Among the individuals most apt to be attracted by this view, especially in its extreme form, are retired middle-class persons, who are homeowners living on a very modest fixed income.[2] Squeezed by inflation and rising taxes, it is not surprising if they fail to see the need for innovation, especially when it means to them an absolute reduction in living standards. The marginal homeowners, the persons who can just barely afford the home they are buying, whatever its price, are also likely to find the caretaker image attractive. They must justify a low-tax policy.

Finally, the fourth image sees the purpose of local government as *arbitrating among conflicting interests.* Emphasis is placed upon the process rather than the substance of governmental action. Although the possibility of a "community good" may be formally recognized, actually all such claims are reduced to the level of interests. The view is realistic in the

---

[1] Arthur J. Vidich and Joseph Bensman, *Small Town in Mass Society* (Princeton: Princeton University Press, 1958), Chap. V, VII; esp. pp. 113–114.
[2] See Amos H. Hawley and Basil G. Zimmer, "Resistance to Unification in a Metropolitan Community," in Morris Janowitz, ed., *Community Political Systems* (Glencoe, Illinois: The Free Press, 1961), pp. 173–175.

popular sense in that it assumes "what's good for someone always hurts someone else." Given this assumption, government must provide a continuous arbitration system under which public policy is never regarded as being in final equilibrium. The formal structure of government must not be subordinated to a specific substantive purpose; rather the structure must be such that most interests may be at least considered by the decision-makers.

That the proper role of government is to serve as an arbiter is held most logically by interest groups that fall short of complete political control, including, especially, the more self-conscious minorities. The numerical or psychic majority does not have to settle for a process but can act directly in terms of substantive conceptions of the community good. The minority can only hope for access.

The most conspicuous interests that self-consciously espouse an arbiter function for government are neighborhood and welfare-oriented groups. Ethnic blocs reaching for a higher rung on the political ladder, and home owners and businessmen with stakes in a particular neighborhood, especially one threatened by undesired changes, make claims for special representation. The psychological minorities—persons low on a socioeconomic scale—also stress the need for personal access. In the four cities it was found that welfare matters were high among the subjects about which city councilmen were personally contacted. This was true even though welfare was entirely a county function. Welfare is a highly personal problem, at least from the viewpoint of the person in need. Perhaps personal access is desired because welfare, received in exchange for political support, is viewed as more legitimate—more nearly "earned"—than that which is received by merely qualifying under bureaucratically defined standards.

Government-as-arbiter does not mean a neutral agency (government) balancing diverse pressures with mathematical precision. The accommodation envisioned is not one that operates according to a fixed standard of equity or political weighting. Rather it is a government of men, each with weaknesses and preferences. It is just this human element in government which opens up the possibility that the imbalance between the majority and the minorities may become redressed.

The four types of images may be classified into two groups: those implying a unitary conception of the public "good" and those implying a pluralistic conception. The unitary conceptions are stated in substantive terms: i.e., "promoting economic growth" and "providing or securing life's amenities." The types which incorporate a pluralistic notion of the "good" stress a procedural role for government. "Arbitration among conflicting interests" clearly denotes this procedural emphasis.

The third type, "maintaining traditional services," presents a special problem with respect to the unitary-pluralistic dichotomy. A policy favoring a limited role for government seems to constitute a unitary conception of the good; however, the premise of this view is that the public good is achieved by maximizing the opportunity for individuals to pursue their private goals. The role of government is held in check so that a greater range of choices may be retained by individuals. Hence, this third type is essentially pluralistic in its objectives.

This unitary-pluralistic distinction is important to an understanding of the relationship between attitudes toward governmental structure and those toward the proper role of government. The formal structure most consistent with arbiter and caretaker government is one which includes a plural executive, ward elections for councilmen, and other decentralizing devices. These arrangements provide multiple access to policy-makers and, by decentralizing leadership, make programmatic political action more difficult.

Centralization and professionalization of the bureaucracy are attractive to those who can control and who desire to exploit control to achieve specific substantive goals. While it is true that the early advocates of civic reform acted in the name of economy, most successful reforms have been followed by action programs. Promotion of economic growth or increased amenities have formed the inarticulate premise of most of these efforts, even though the more neutral language of procedural adjustments has often been employed. Centralization, through strengthening the office of mayor, the appointment of city managers, and the institution of at large elections, usually leads to a reduction in the political strength of certain particularistic interest groups or minorities. Professionalization of the personnel in a city bureau may achieve some economies, but it also creates a political force pressing for the extension of the services performed by the bureau. Thus the reform devices are most consistent with the two types of unitary views of the commmunity good.

It should be stressed at this point that the prime concern here is with the development of the typology and not with the implicit hypotheses regarding the groups which support each type. The foregoing references to group interests are used only to suggest the plausibility of the types by citing the relationships among them and familiar forces within communities.

The typology has been elaborated in terms of images of the proper role of local government. The necessary data for verification of the types would include some kind of opinion sampling. However, the typology may be used for other purposes and differing kinds of data. It may be converted into a typology of *actual* roles of government or even prevailing norms among all community policy-making institutions.

This study uses the typology in the second sense: as a classification for the actual role of government. For this purpose the emphasis in data collection must pertain to prevailing policies. For a policy typology, the fourth category, arbitration between conflicting interests, poses a problem. Arbitration is not, strictly speaking, a policy in the same sense as promoting economic growth, providing and securing life's amenities, or maintaining traditional services. However, in classifying prevailing governmental policies this fourth type may act as a useful category for characterizing cities in which the policy complexes embodied in the other three types are continuously compromised and no one or even pair of the other policy types epitomizes civil actions over any substantial period. However, in some cities the arbiter type, as first defined, will be useful as a completely parallel policy category. This will be where the process of adjusting claims among various and disparate interests becomes the all-pervasive preoccupation of officials and the norm for policy determination generally.

In the literature of case studies in community decision-making there are illustrations of each of the four types. Springdale village, described by Vidich and Bensman, is nearly a complete prototype of caretaker government.[3] Whyte reports the preoccupation with amenities in the politics of Park Forest.[4] The government of Middletown as described by the Lynds was primarily concerned with the interests of the producer.[5] Perhaps the best illustration of arbiter government is found in Meyerson and Banfield's study of public housing site choices in Chicago;[6] 'their study pertains to only one governmental policy, and is, therefore, not an accurate characterization of Chicago government generally. But with regard to policy formation in this one area, Meyerson and Banfield stress the way in which the political party, working through the city government, acted as an arbiter compromising the aims of all claimants in the political struggle.

APPLICATION OF THE TYPOLOGY

The usefulness of the typology is dependent upon developing satisfactory criteria, which provide the basis for distinguishing the policies of the various cities. As the very character of the types discourages the use of any rigorous mathematical measurement, primary reliance must be placed on

[3] Vidich and Bensman, op. cit.
[4] William H. Whyte, Jr., The Organization Man (New York: Simon & Schuster, 1956).
[5] Robert S. and Helen M. Lynd, Middletown in Transition (New York: Harcourt, Brace & Co., 1937).
[6] Martin Meyerson and Edward Banfield, Politics, Planning and the Public Interest (Glencoe: The Free Press, 1955).

descriptive material. In the final summary an attempt is made to assess comparatively the content of municipal politics, according to the subtle factors as well as the more obvious ones set forth in the above description of the types. However, primary emphasis will be on the following criteria.

A commitment to *economic growth* was gauged primarily by two standards: (1) policies designed to recruit industries or business that would increase employment or the total wealth of the community, and (2) policies that endeavored to preserve the fruits of growth for core-city interests. The problems attached to industrial recruitment will not be elaborated here, but under certain circumstances they involve annexation, utility expansion, zoning policies, and tax rates. Efforts to preserve the fruits of growth usually related to policies that favored central business interests as opposed to commercial developments on the city fringes.

An assessment of a city's interest in providing life's amenities was made according to the range and the levels of services provided. A fuller range of services represented a greater commitment to this standard. Evaluations of professionals were relied upon to determine service levels in many functional areas where city policies are related to citizens as consumers. It was assumed that a city which professionals judged to have superior service would rank high in terms of the amenities on the ground that professionals generally favor increased emphasis on their particular specialties.

The *caretaker* values were defined by an emphasis on private as opposed to public allocation of goods through the maintenance of only traditional services and, consequently, a low tax rate.

Arbiter government implies freedom of access for all groups and specifically for minorities. Access is maximized when the city councilmen are willing to act as delegates for particular constituencies. Thus the measure of arbiter government rested largely on whether a city council was dominated by individuals who became advocates for various claimants or by persons who rejected the legitimacy of this conception of representation and guided their decisions by a unitary conception of the community's interests.

While the possibility of a rigorous quantitative scale has been denied, at least, with the data to be presented, a scale of sorts is developed. Consequently, in the interest of candor, as well as a means of presenting a brief visual capitulation, Figure 1 is offered. In it four Michigan cities have been placed on a subjectively derived scale showing gradations of commitment to each of the four policy types. In doing so the universe of comparison is confined to the four cities. Undoubtedly, there are other cities which would scale higher and lower in each category. Further, each set of values is considered independently of the others. An amalgamation of the separate elements will be attempted next.

	Low	→	Medium	→	High
Economic Growth	Delta			Beta	Alpha
				Gamma	
Life's Amenities	Delta		Beta	Gamma	Alpha
Caretaker	Alpha	Gamma		Beta	Delta
Arbiter	Alpha	Delta		Beta	Gamma

FIGURE 1   *A typological profile of the four cities.*

Several questions may occur as a result of this chart. (1) Are some of the types merely the reciprocal of others and therefore not truly distinct types? (2) If not reciprocals, are not some pairs more apt to be incompatible than are others? (3) If the categories are truly distinct, is not the chart inconsistent in showing a given city high on the scales of two or more mutually exclusive sets of values?

The answers to all these questions lie in the fact that the types are empirically, not logically, derived, even though the underlying assumptions of each of the types are logically distinct. The confusion stems from the fact that in actual behavior it is possible for a city simultaneously to follow contradictory policies. Only under certain conditions of duress are such conflicts forced to the surface. Thus, we would assume that none of the types is a complete reciprocal. For example, even though a city's leaders may place a low priority on amenities, they will not necessarily be committed to a caretaker government. This implies the answer to the second question. Within the experience of these four cities, there appears to be a pattern of disparity between certain pairs of types. It certainly seemed so for the caretaker and amenities types. However, it is conceivable to have nearly any possible combination. The answer to the third question has also been implied. Policies contradictory with regard to underlying principles are not necessarily perceived as contradictory in an actual political setting. Furthermore, in any given universe of a limited number of examples, cities may be found that stress both of two distinctly different sets of values in comparison with other cities. Our experience with the four cities indicated that sufficient slack existed between areas of possible choice for most cities to have the luxury of pursuing quite diverse goals.

There is one further apparent difficulty associated with applying each of the four types separately to the four cities. Preferences are shown only in relationship to choices in the other cities. Alpha is rated high in terms of amenities and low in terms of caretaker principles because other cities, especially Delta, rate in the opposite fashion. Similarly, Alpha rates high in terms of both amenities and economic growth because no other city rated higher. As a result there is no indication as to how Alpha behaved when the values of consumers clashed with those of producers.

To a certain extent this difficulty is a function of the fact that our universe is small. If we had some measure of value preferences in all American cities, our scale would expand, revealing differences more precisely in a comparative fashion. There probably are prototype cities which subordinate most policy to a single standard. The company towns, the upper middle-class suburbs, and the isolated village provide examples where policy is dominated by producer interests, consumer interests, or *status quo* interests. Inclusion of such varied places would give greater polarity to a scale.

. . . .

SOME EMPIRICAL FINDINGS

Helen and Robert Lynd suggested that the difference between local and national government is the difference between concern with how income is earned and how income is spent. Obviously this is an over-simplification, but it may suggest a distinction helpful in explaining the differences in political behavior reported here. All levels of government are involved in policies having implications for both producers and consumers. However, even in an age of an emerging social service state, local government is more a consumer-service-oriented agency than are government systems at the higher levels. It is less involved in regulation than in providing consumer commodities in a socialized fashion: commodities such as water, sewerage, streets, fire and police protection, parks and health services. While all citizens consume these services and may desire more of them, the wealthy individuals are more capable of paying for them and appear to be more desirous of having them. City services are a necessary part of their style of life. As a result, these persons exhibit concern regarding the administration of them. The data on voting responses on tax referendums lend support to this theory.

Our data also indicate that the consumption-oriented upper-middle class is often also favorably disposed toward policies of economic growth. Economic growth and amenities, when stated in general terms, can be made to appear compatible. "The city which is an attractive place to live is also the city attractive for new industries." But on specific issues, the values may clash as they affect individual citizens. In Gamma the choice had to be made between the two when a chemical manufacturing firm wanted a site adjacent to a residential neighborhood. Alpha made a choice between a renewal project and downtown revitalization. But in these middle-sized cities only occasionally were the two sets of values thrown into direct competition. More commonly, the choice was whether or not to initiate a specific measure, such as municipal parking or creation of a new park. When raised in an *ad hoc* fashion, the decision was likely to turn on the momentary strategic position of particular interested parties.

While both economic growth and increased amenities were frequently political expressions of the more well to do ranged against the caretaker sentiments of the less well to do, this does not furnish us a theory to underpin the typology. There was little indication that the distribution of persons in the economic levels varied significantly from one city to another.

COMMUNITY CHARACTERISTICS

The relative distribution of income levels did not provide a sufficient explanation, but variations in the economic structures of the communities proved more fruitful. Alpha exhibited the most steadfast adherence to policies that have been identified with the preferences of higher-income citizens. It emphatically rejected caretaker government. The most distinctive feature of the economic structure of this community was its higher percentage of home-owned industries. It was the home office of several large national firms. It had many other small but substantial locally owned industries. The founding families were still prominent in the management of most of these industries, with the Brinton-Lowe family the leading example. In this respect Alpha was a survival from the American past, a past where local indigenous capital, rather than national capital, controlled industry and executives did not look longingly toward life in some distant metropolis. If the executives wanted to control their living environment they had to do it in Alpha. They were not persons with a "limited liability" in the community. Consequently, management was willing to free top talent for local politics. Shortly after the close of our study period, one of the leading sons of the Brinton-Lowe family was elected mayor. Upon the occasion, he resigned a position of prominence in his firm, announcing that he would henceforth give full time to civic affairs. Thus Alpha had a "fifty-thousand-dollar" executive serving as mayor—for one thousand dollars.

The longevity of local ownership may have in turn resulted from the stability of these local industries. The two industrial mainstays, paper and pharmaceuticals, were comparatively depression-proof. Mergers for purposes of diversification were not an essential for these industries, as there was no compelling necessity to seek stabilizers to cushion against fluctuating economic fortunes. Wages were not especially high in these industries, but employment was stable. During the Great Depression when other cities continually cut back capital improvements or plunged into debt, Alpha was able to maintain its municipal plant and at the same time maintain a debt-free city. Partly as a function of its economic position and partly as a result of a professional administration Alpha provided services compatible with upper-middle-class consumption demands and, at the same time, did not tax the lower-income groups severely. (Alpha also benefited from a legal arrangement in the immediate postwar inflationary period. From the end of the war until 1949, Beta and Gamma operated under a statutory low-tax ceiling forcing them to cease all capital outlays. Alpha was not subject to this ceiling, which could only be removed by the state legislature.)

In the analysis of the referendums, it was seen that Alpha drew broad support for civic improvements from all but the very poorest sections of the city. A sewage bond was ratified by 86 per cent of those voting. While the Citizens' Committee political organization was undoubtedly one factor in evoking this response, the economic stability of Alpha would seem to be another. The working people of Alpha had no legacy of experience with prolonged unemployment during which the property tax had pressed heavily on home purchasers. Security of employment at least is a reasonable hypothesis to explain the broad support reaching into lower income groups.

Beta, Gamma, and Delta had similar economic structures and a bulk of the industries were, or were rapidly becoming, absentee-owned. These same industries were more sensitive to national economic fluctuations. Heavy concentrations in durable goods created a need on the part of management to merge with other firms. Public apprehension concerning the future was engrained into the local culture. Gamma and Beta had seen many large industries leave and fewer enter. Delta had not lost any major industries, but neither had it gained any, and the existing larger industries had widely fluctuating labor forces. The average incomes for the four communities appeared to be the same, but the average hid the fact that Alpha had lower, though stable, wage rates; in the three durable goods centers, incomes for those employed may have been higher, but many did not work a full week, or did not work at all.

The old economic elites of Beta, Gamma, and Delta were gone in the flesh and remained only enshrined in the stone of philanthropic monuments erected in the past. In Gamma, many public buildings and parks were decorated with the names of old lumber barons, largely extinct by the turn of the century and completely gone as the Great Depression began. Beta sported a large park, the product of similar philanthropy in the 1920's. Interestingly, hardly an additional acre had been turned to park use since and, as a result, the recreational facilities remained disproportionately piled in the wealthy corner of the city, where residents already possessed parklike lawns and backyards.

Unfortunately for the sake of research, none of the four cities was both economically stable and dominated by industry with absentee ownership. The problem for our analysis is whether the more important factor is the ownership pattern or mere economic prosperity. A second problem is posed by Gamma. While this city suffered all the economic privations of Beta and Delta, it consistently rated higher with respect to amenities and did not display the same intransigence on tax referendums.

Besides variations in the economic structure, gross community differences were also present with respect to ethnic composition. The source of difference may have contributed in a small way to an explanation of Gamma. Beta and Delta historically and contemporarily possessed large ethnic blocs, internally more cohesive than those in Alpha and Gamma. The most important of these groups was the Poles, a bloc which has been credited with retaining a certain separateness by Thomas and Znaniecki.[7] Though we are dealing with a later generation, some of these characteristics continued to characterize the political activities of the Poles in these two cities. In neither city were they a numerical majority; however, their influence was in part a legacy from past actions, which saw the creation of ward systems. In Delta, this effect can be traced to a specific incident: in the 1920's when the Poles, in conjunction with the French-Canadian and German neighborhood blocs, successfully worked for the reinstatement of a ward system after a brief trial with at-large elections. In the 1950's these same neighborhoods contributed heavily to the defeat of a new proposal for at-large elections.

In both cities the Polish population was largely working class and lower-middle class. None of its members had penetrated the economic elites of the cities. One found no Polish bank officers, editors, industrial executives, or prominent professional men. There were a few successful entrepreneurs, especially in retailing and construction. In both cities, frequently the losing mayoralty candidate belonged to a major ethnic bloc, usually Polish. In both cities, but especially in Beta, the Polish neighborhoods gave a constant negative vote on proposals for civic improvements. Thus, there is some evidence that these ethnic blocs, especially the Polish ones, were still concerned with a politics of status. This appeared to affect

---

[7] William I. Thomas and Florian Znaniecki, *The Polish Peasant in Europe and America* (Chicago: University of Chicago Press, 1918), Vol. I, Introduction, esp. pp. 152–156.

their political behavior with regard to traditional political issues of city government. They represented a public difficult to mobilize for the "broader" community aims implicit in policies directed toward economic growth and expanded amenities.

The most conspicuous ethnic groups in Alpha and Gamma were the Dutch. The initial arrival of the Hollanders preceded that of the Poles in Delta and Beta by about a generation. Either as a result of this earlier arrival or because of a different cultural outlook, the Dutch entered into the civic life more as individuals than as ethnic spokesmen. Many became prominent among the economic elites. Indeed, it was part of the public lore in Alpha that the local civic pride and concern for amenities were fostered by Dutch cultural traditions. "Why is Holland so clean? Because everyone sweeps his own doorstep." This riddle was often heard in Alpha as an explanation of the local civic virtues.

To the extent that the ethnic blocs of Beta and Delta helped to maintain a ward system and to stress a politics of neighborhood interests, they contributed to a political system that recruited and strengthened the forces within the cities which were attracted by caretaker government and were often hostile to policies of economic growth and increased amenities.

THE POLITICAL PROCESS

Political institutions may reflect stable value preferences in a community and thereby make the form of the decision process merely a function of dominant values. However, political forms and traditions may develop to fit one set of circumstances and at a later time their very existence may influence the course of decisions in a manner unanticipated. A manager may be chosen for his traditional managerial skills, but once in office he may act as a rallying point to further goals that were not considered at the time of his selection. A charter commission may choose ward elections to achieve "well-rounded representation," but the consequences of this choice may be to favor a particular view. Once the ward elections are in use the redistribution of strategic positions in the political process may prevent a recall of the decision. In this sense, variations in the composition of the political process may be an important factor in explaining policy variations and may not necessarily be simple reflections of dominant value preferences in the communities. In this section, the character of the municipal bureaucracy, the council-manager form, nonpartisanship, and electoral arrangements will be considered in this light.

A major element that distinguished the governments of Alpha and Gamma from those of Beta and Delta was the strong professional traditions in administration existing in the former. The presence of a structured politics gave strong support to the bureaucracy of Alpha, but in Gamma the bureaucracy was more exposed and had to seek its own support. However, the manager was a long-time incumbent and had earlier served as manager of one of the satellite suburbs for nearly a decade. The manager of Gamma combined great political knowledge of the community along with his professional skills. During the years when the council was dominated by men who assumed a broker's approach to their jobs, the manager was under constant attack because his values were directed toward goals associated with economic growth and amenities. Although the council could place obstacles in the way of the manager's program and even control areas of personnel policy, it never succeeded in dismissing him. Elections would be held, the manager would receive a more favorable council, and his program would continue. For instance, the manager kept plugging away for over a decade for the construction of a major four-lane, limited access highway into the very heart of the city. In this case, his accomplishment was a function of his knowledge of state politics as well as of local conditions.

By contrast, Beta had a tradition of short-term managers. During the latter half of the study period a professional manager arrived on the scene who initially experienced many of the defeats by the caretaker elements that previous incumbents had also experienced. After approximately five years of tenure the manager began to achieve more of his goals. Gradually department heads were replaced and a professional staff was assembled. The developments subsequent to our study period indicate that if the typology were applied now, Beta would have to be placed in a different category. The question is: Did the manager create the change or did the political climate of the community change? There were changes emerging in the political organization of the city. The weak old Association for Sound Government expired and a new businessmen's group began covertly recruiting and electing councilmen. The new recruitment method showed some promise of affording organized support for the manager. But a manager who was experienced in the community and who had a program for change arrived on the scene before formation of the new organization. This suggests, but does not prove, that this latest manager was the significant factor in explaining community change. It must be added, even though it only complicates analysis, that professional managers could not survive in Delta long enough to develop personal sources of support. Thus, our four cases indicate that a manager is only one force in remodeling policy even where the political process is not already highly structured. On the other hand, the existence of a strong manager may obviate the necessity for groups agreeing with the manager's policies to organize. They never did in Gamma, except sporadically when the manager was in trouble.

It has been implied in the foregoing that the council-manager form of government is not neutral regarding city policies, that it favors one type of policy over another. Considering the importance of this subject to the students of local government, perhaps it deserves additional comment. In the first chapter three basic assumptions underlying the council-manager plan were set forth. These were: (1) that the council should represent the city as whole, rather than discrete interests, (2) that administration should be conducted by professionals, and (3) that the council should control policy; the manager, administration. It is not our intention to disinter the arguments in public administration that have led to drastic qualification of the third point. But while the politics-administration dichotomy is now poor public administration theory, it contains an element of practical wisdom. Probably the language used by early administration theorists was too stark, but nevertheless the council-manager plan does require a certain mutually satisfactory delineation of functions between the council and manager. The line is not between policy and administration, but rather between policy supported by community consensus and policy that has not yet achieved this consensus. Furthermore, if there is not some stability in the content of these mutually agreed-upon spheres, the manager will experience difficulties. This is one reason for the first principle, which requires at-large elections. In the terminology of this study, the manager plan is incompatible with a brokerage approach to representation. Of the four types, arbiter government would seem to be the least likely to produce an adequate framework for mutual understanding between professional administrator and lay councilman.

Arbiter government is characterized by the absence of a fixed standard in allocating city resources and favors. If each councilman is a delegate of neighborhood interests, or acts as a channel of access for shifting interests and disparate groups, the manager is robbed of a standard by which to guide his administration. Pressure will continually be placed on the manager to fix *that* street in *that* ward. If services or policy is dispensed to obtain political support, the council, not the manager, is the source of allocations. If allocations are on a rough rotation system, as was the case in Delta, the skills of the manager are greatly downgraded.

But in Gamma, and to a lesser extent in Beta, the manager retained an important administrative and policy role, despite the degree of arbiter government, because the sanctions of the individual councilmen were not so great. There was never any organization that could completely challenge the manager and his professional standards. Even so, the manager was at times administering policies that he felt were undesirable. But he never faced a political organization that looked to the elections as a time of crisis. There was always the prospect of change in councilmanic membership if he bided his time.

The other three types of policies—amenities, economic growth, and the maintenance of traditional functions—ostensibly provide a consistent basis for action. However, caretaker government is in fact, if not in theory, even more in conflict with the council-manager plan than is arbiter government. It must first be recalled that the city manager considers himself a member of a profession. He is a person skilled in the art of doing. But the manager is not simply an administrator, for his profession incorporates a sense of mission. The city manager's code of ethics asserts that "The city manager . . . has a constructive, creative, and practical attitude toward urban problems and a deep sense of his own social responsibility as a trusted public servant."[8] Thus the manager is a problem-solver by trade. It is against his professional code of ethics to let the city's physical plant deteriorate for the sake of low taxes.

The clash between the manager plan and caretaker government does not stop with professional values, however. Career advancements for managers are based upon concrete achievements, not simply satisfied councilmen. The profession-oriented manager is outward-looking toward his peers in the International City Manager's Association. He may sometimes be more interested in their opinions of him than he is in those of his councilmen. Their judgments, rather than those of local citizens, tell the manager whether he is a success or failure. In a study of city manager attitudes it has been reported that managers believe they should do things that will enable themselves to "move up" to more desirable cities. Ability to move is an indication of professional status. Any action on the part of the city council to rob a manager of the opportunity for mobility should logically prompt the manager to resign.[9] Beta and Delta, the two cities that embraced caretaker principles, also had the greatest difficulty in retaining city managers. Indeed Delta could not (or would not) retain anyone with professional experience. The managers who did remain in office for several years were ex-politicians who had risen through the patronage system. These Delta managers had no potential for mobility in the professions. Their managerial careers began and ended in that city.

The amenities and economic-growth types are much more appealing to city managers, for these concepts of government provide a basis for manager initiative. Furthermore, the likelihood of manager-council clashes is reduced under them, because administrative policies have guidelines.

[8] From The City Manager Code of Ethics, as revised in 1952.
[9] See George K. Floro, "Continuity in City-Manager Careers," *American Journal of Sociology,* **61,** 240–246 (1955).

In addition, managers have greater opportunity to further themselves professionally. Civic monuments reflect to their credit in the form of storm-sewer systems, industrial parks, shaded thoroughfares, and pedestrian malls. It is not surprising to find that in Alpha and Gamma the tenure of the city managers was usually long and the incumbents were men highly regarded in the profession. These two cities illustrate the kind of city toward which managers aspire. Alpha, especially, was one of the prized assignments.

Increased amenities and economic growth provided a congenial policy framework for the ambitions of individual managers and simultaneously the goals of such values demanded professional leadership. While the council-manager plan is not the only way of obtaining this aid, it is a convenient and familiar alternative.

Similarly, the question may be asked whether nonpartisanship favors any particular type of policy orientation. The answer is a qualified yes. Our study has indicated that fewer people on the lower end of the socio-economic scale vote on the nonpartisan than on the partisan ballot. Even when partisan and nonpartisan elections were held simultaneously, as was the case in Gamma, this was true. An exception to the pattern appeared in Delta, where the union organization apparently stimulated the turnout of lower-income persons. The inquiry into tax referendums and urban renewal suggested that caretaker government had broad support in lower-income areas. Thus, to the extent that the nonpartisan ballot reduces voting among lower-income persons, it favors the achievement of policies concerned with amenities and economic growth. One may only wonder what the consequences would have been in Alpha, Gamma, or even Beta if the voting turnout in low-income precincts had been matched by the record of similar precincts in Delta. Perhaps this is another way of saying that as long as civic improvements are based upon the *ad valorem* property tax, they will be achieved only because of self-disenfranchisement of a large number of tax-sensitive citizens; conversely, political organizations drawing support from lower-income groups will not embrace economic growth or improvement in amenities as the primary function for local government.

Regarding at-large and ward elections, there was a parallel between the kind of representation and the policy orientation of the cities. The fortunes of the Delta union organization under at-large conditions may have differed substantially. The conspicuous failure of the union in mayoralty elections was probably only partially a function of union interest in the office. The whole political organization and its techniques were geared to the scale of small wards. It is equally difficult to picture the Citizens' Committee of Alpha maintaining its hegemony under the ward system. The bulk of its councilmanic recruits were drawn from three precincts out of a total of thirty-two to forty. If the Committee had to work through pawns recruited from various sections of the city, it is doubtful if the council would have enjoyed the same kind of civic repute as did the existing councilmen.

Again, however, the ward versus at-large question is one factor among many. Gamma and Alpha both had at-large systems. Both did shy away from caretaker policies. But this is a very small sample on which to build bold assertions. Beta and Delta had ward systems and both were lower on the scale in regard to the provision of local amenities. Perhaps the strongest statement that can be extracted from this evidence is that the claims civic reformers have made over the past half-century are not contradicted by the data in this study. At-large elections are a necessary but not a sufficient condition for enabling upper-middle-class values to prevail. Similarly, the objections of labor union leaders to at-large representation appear valid, given their value assumptions. In Gamma, even with sustained action, the labor unions could gain no more than one representative. In Delta, ward organization brought control of the council. However, without organization there is no guarantee that the ward system would produce victories for working-class interests. To be sure, Beta had a partial ward system, and the prevailing low-tax policy probably pleased the low-income families. But this policy was probably not a self-conscious recognition of the interests of the less well-to-do citizens.

The success of a policy very likely will have a reciprocal effect on other policies. The ability of Alpha to annex large suburban areas dramatized the similarity between the consumer-service orientation of that core city's policies and those which are frequently associated with suburban areas. (This does not imply that residential suburbs are able to afford such policies, but only that amenities are the preferred goals.) Alpha's political style, dominated as it was by business and university leadership, avoided the core-city stigma of lower-class status, poor amenities, and raucous disputes. Thus, the ability to annex probably meant that the political traditions would be maintained through the transfusion into the political system of new citizens likely to support the Citizens' Committee tradition. The disparity between the political values of Beta and Delta, on the one hand, and their suburbs on the other, precluded similar annexation in those cities. (Gamma's special metropolitan problems made this test largely irrelevant there.)

SUMMARY

Our data have yielded some insights into the factors underlying preferences for the first three types of policy orientations, but little has been said about the fourth. The difficulty is that the political structures in these communities long predated our study period; the origins of the Citizens' Committee and the Delta union organization lie buried in the past. Our comments on this subject must be confined to the implications of structured politics for the political process generally. This will be given attention in the last chapter. The following generalizations are borne out by the data about the first three types:

1. To the extent that policies of economic growth and amenities call for increased expenditures, support generally comes from the higher income groups and opposition comes from lower income groups.

2. Conversely, the strongest supporters of caretaker government are centered among low-income citizens.

    a. To the extent that nonpartisanship reduces the working-class vote, it also weakens the forces for caretaker government.

    b. Contrarily, ward elections strengthen caretaker policies.

    c. The existence of cohesive ethnic blocs generally adds to the political strength of working-class citizens and thereby strengthens the forces for caretaker government.

3. Professional city managers favor values associated with both economic growth and amenities.

    a. The effectuation of manager values are associated with long tenure.

4. The economic climate of a city is insufficient in itself to determine the value orientation of civic policies.

5. High incidence of home-owned industries is correlated with high rating in the scale of preferences for amenities and economic growth.

6. Caretaker values are more in direct opposition to policies of increased amenities than are those providing for economic growth, for the latter are less frequently associated with costly expenditures.

7. In middle-sized cities, producer-consumer conflicts are rare.

ROBERT EYESTONE

HEINZ EULAU

City Councils
and Policy Outcomes:
Developmental Profiles

STATEMENT OF THE PROBLEM

As we travel, with our eyes open, through the many cities large and small of any metropolitan region, we are struck almost immediately by certain similarities and differences in their physical and social appearance. In one city the traffic flows smoothly and parking space seems abundant, while in another city (of equal size) there is traffic congestion and pedestrian turmoil. In one city, clearly affluent by the looks of its homes, there are no sidewalks, street signs, or protected crossings; in another city, equally affluent, public convenience and safety are well taken care of. In one city we notice a swimming pool, a library, and perhaps a museum; in another city, otherwise similar in appearance, these amenities of urban life are missing. In one city residential and commercial areas interpenetrate and seem to defy any order, with fringe growths that are monuments to ugliness; in another city, also balanced, the areas devoted to work and home life are well separated, land use seems satisfactorily controlled,

The larger project of which this analysis is a part, the City Council Research Project, is sponsored by the Institute of Political Studies, Stanford University and is supported by the National Science Foundation under contract GS 496.

and the city's fringe does not serve as a dumping ground. In one city the streets are full of children, idle or playing randomly; in another city the children populate attractive playgrounds and games are organized or supervised.

These are only some of the more obvious things we see on casual inspection. Hidden behind these outward appearances are many other differences in what we may call a city's "public life style." Yet, it is an ordinary observation that the cities in a metropolitan area face many common challenges that stem from the common environment. In a rapidly growing region such as the San Francisco Bay Area where our research is being conducted, population growth and attendant changes in the region's ecological and socioeconomic character are necessarily felt in all cities, though the resultant pressures may differ, depending on a city's history or location. Yet, even granted that the common metropolitan problems have a differential impact, some cities seem to respond in one way and other cities in another to environmental challenges. Common pressures from the environment apparently can be interpreted differently in the process of converting them into public policies.[1]

How can we account for the fact that cities facing basically the same challenges from the environment seem to react so differently that it is possible to see distinct differences in their public life styles? The conventional answer is to point to a city's socioeconomic character—whether it is industrialized or not, commercial or not, residential or not, balanced in all these respects or not. But we shall seek for another answer by asking this question: could it be that differences in meeting the common environmental challenges, giving rise to different public life styles, are due to the fact that different cities are in different stages of urban policy development? Once we ask this question, a host of others come to mind: do different stages of policy development correspond to differences in the impact of environmental challenges? How can such stages be identified? Do they follow each other with linear regularity, or are reversals in policy development possible? What makes for "retarded" development in one city, while a similar city "takes off" and a third "matures?" What is the character of the policies designed to meet environmental pressure? Do they represent attempts to adjust the city to the changing environment, to control the environment, or both?

We may ask another set of questions: if cities are in different stages of development, is it the result of differences in resource capabilities by which environmental challenges can be met? But how is it that cities

---

[1] See, for example, the recent work by Oliver P. Williams et al., *Suburban Differences and Metropolitan Policies* (Philadelphia: University of Pennsylvania Press, 1965).

with more or less equal resource capabilities may yet respond differently in terms of policy development? Are policies formulated by public policy-makers in response to demands for services or amenities by interested individuals or groups? What role do the decision-makers' own policy preferences or attitudes play in all of this? Does the city's "group life" or its policy-makers' orientations toward development and the scope of governmental activities correspond to the stage of development the city is in?

We do not propose to answer all these questions here. But they point up the problem we wish to investigate and they do provide the cutting edge of a theory of urban public policy. They also suggest at least some of the variables that should be formalized in a model of urban policy development as well as some very specific questions that we can ask of empirical research.

Before presenting a model of urban policy development and testing some hypotheses derived from that model, we shall explicate the concepts we are using and describe their empirical measures.

CONCEPTS AND MEASURES

Policy, Policy Outcomes, and Policy Profiles

The *policy* of a governmental unit is the total relationship of government to its environment, as expressed in its concrete programs and specific decisions. We conceive of public policy as a response of government to challenges or pressures from the physical and social environment—wars, famines, depressions, population pressures, technological advances, clashing interests, and so on. Changes in public policy, then, can occur in response to changes in the environment. The response can be twofold: either the policy adjusts and adapts the political system to environmental changes, or it brings about changes in the environment. Which alternative is chosen, or whether both are chosen, depends potentially on a great variety of factors—the structure of the political system, including the vitality and diversity of its group life; the functions which it is seen to perform traditionally; its resource capabilities; and the values that policy-makers seek in formulating policy—their policy orientations.

The problematics of policy-making arise out of the relationship between changes in the environment that are experienced as challenges and require some kind of response, on the one hand, and the policy orientations of decision-makers, on the other hand. For if policy is a response to environmental pressures, both physical and social, as well as an anticipation of a future state of affairs, a change in policy is both causal and teleological. It is "caused" by environmental challenges, but it is also directed toward an end in view or shaped by a purpose.

It is the tension arising out of the interchange between causal and teleological "forcings" that makes public policy so tantalizing an object of scientific investigation. Moreover, policy can be a response to the environment without particular decisions having been made. In other words, no government can be without a policy, even if government officials refuse or fail to make decisions regarding a policy to be adopted. Their very failure to decide and its consequences for governmental functions constitutes a policy in the sense that it is a response, if a negative one, to the environmental challenges facing them.[2]

Because a "no-policy" is also a policy, the study of public policy cannot rely on manifest statements or overt decisions exclusively, but it must also concern itself with policy outcomes as the most direct evidence of how environmental challenges have been met. By policy outcomes we mean the tangible manifestations of policy—revenues, expenditures, regulations, administration of justice, application of sanctioned violence, and so on. Policy outcomes, then, reflect the action orientations of policymakers, regardless of whether a decision has been made or not.

The study of policy outcomes is predicated in a classification of policies. Such a classification should accommodate categories and empirical indicators appropriate to what we have described as the adaptive and control aspects of policy. As, in the present research, we are dealing with policy outcomes in cities, we will be using expenditures for planning as an indicator of the "control function" and expenditures for amenities as an indicator of the "adaptive function."[3]

The measure for the *adaptive function* is the percentage of total government expenses spent for health, libraries, parks, and recreation. These categories are used partly for convenience, since they are major accounting categories used to report city expenditures, although they presumably include the major amenities offered by cities. A high amenities city would contrast with a city with a traditional services orientation, which would spend a high proportion of city income for fire and police services or public works.[4]

---

[2] See Peter Bachrach and Morton S. Baratz, "Decisions and Nondecisions: An Analytical Framework," *American Political Science Review,* **57,** 632–42 (1963).
[3] Since education and public welfare policies are not made at the city level in California we could not use expenditures in these areas as measures of policy outcomes in our study.
[4] The amenities measure is an attempt to tap Williams' and Adrian's concept of

The measure for the *control function* is the percentage of total general government expenses spent by the planning commission. General government expenses include essentially all administrative expenses and salaries not included under fire, police, or recreation categories, and so on. Expenses by the planning commission include both expenses and outlays, therefore encompassing the range of items from paper supplies to salaries of full-time city planners to special outside studies commissioned by the city planning commission.

California State law requires every city to have a planning commission, but this body may be, and frequently is, a standing committee of citizens appointed by the city council and incurring no expenses charged against the city. Therefore, the actual dollar amount spent by the planning commission would seem to be a good indicator of the extent of a city's commitment to the idea of planning as a way to control the environment. General government expenses were used as the percentage base rather than total government expenses in order to make planning definitionally independent of amenity expenditures.

In order to be able to characterize each city in each year over a five-year period (1960 to 1964) as either "high-spending" or "low-spending" 'with regard to planning and amenities expenditures, we are using the median of the medians for the five years for all cities as the cutting point. This makes comparison of policy outcomes across cities and through the five-year period manageable.

Policy Development and Developmental Typology

As our interest in policy outcomes is to determine the extent to which they are responses to changes and challenges in the environment, we constructed for each city a five-year *policy profile*.

Policy development is conceived of as a set of policy outcomes that follow each other sequentially through time. If the outcomes are similar, we refer to the sequence as a *stage* in policy development. For instance, the literature on development in the new nations may speak of such stages as "traditional," "transitional," and "modern."[5] In this study, we shall

---

amenities. See Oliver P. Williams and Charles R. Adrian, *Four Cities: A Study in Comparative Policy Making* (Philadelphia: University of Pennsylvania Press, 1963), pp. 198–225.

[5] See, for instance, David E. Apter, *The Politics of Modernization* (Chicago: University of Chicago Press, 1965); or Lucian W. Pye, *Aspect of Political Development* (Boston: Little, Brown, and Company, 1966). Also Robert A. Packenham, "Approaches to the Study of Political Development," *World Politics,* **17,** 108–20 (1964).

locate cities in three developmental stages which we call *retarded, transitional,* and *advanced.*[6]

The conception of a set of sequential outcomes, summarized in a policy profile, as constituting a stage implies continuity and stability. But the notion of development means that one stage, sooner or later, will yield to a new stage. Yet, it is unlikely that a given stage will suddenly be replaced by another. Unfortunately, some of the writings on development that use the concept of stage as an analytical tool give just this impression. Stage A is followed by Stage B and so on. It stands to reason, however, that the developmental process does not neatly subdivide into clear-cut stages that follow each other with unfailing regularity. Leaving aside for the moment the possibility that development may actually be reversed, even if the developmental sequence is "progressive" the transformation from one stage to another may, and perhaps must, involve a sequence of policy outcomes that are dissimilar—some more appropriate to a policy profile characteristic of Stage A and some to Stage B. In this case the direction of the developmental process appears uncertain. Or, put differently, this case gives rise to a policy profile from which we cannot easily predict whether the system will remain in Stage A or be transformed into Stage B.

For instance, to take an extreme example, let us assume that we are observing five policy outcomes that follow each other in an annual sequence. $Outcome_1$ looks as if it fell into a profile from Stage A; $Outcome_2$ however, might look more appropriate for inclusion in a profile from Stage B, as does $Outcome_3$; but it is followed by $Outcomes_{4,5}$ which are characteristic of Stage A again. In other words, in the period of observation the outcomes are sufficiently dissimilar so that the resultant profile cannot be assigned to a single stage of development.

We define a stage as a profile with similar outcomes, and we define a profile composed of dissimilar outcomes as a *phase*. The notion of phase

---

[6] It is important to keep in mind that our observations cover only a small segment of the historical developmental process, a segment that is essentially "modern." We can illustrate this, as follows:

Three Historical Stages of Development

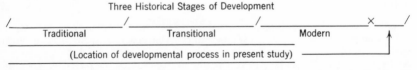

It is all the more significant that, even within this small segment of history, we can locate cities in clearly different stages of policy development. This suggests that a concept like "modern" disguises a great deal of variance that microanalysis can reveal. The point to be made is that our stages "correspond" only analytically to similarly named stages used in the long-term analysis of national development.

suggests that the period in question is less clearly bounded and, presumably, of shorter duration than a stage. Because in a five-year period any city can be in one of three stages of development, we will be dealing with two phases—an *emergent* phase located between the retarded and transitional stages and a *maturing* phase between the transitional and advanced stages.

A word is in order on "reversed" development. While stages of development (as constructed by the historian) are inevitably consecutive and irreversible, as when we say that the Middle Ages "followed" antiquity, stages as we conceive of them for analytical purposes are in fact reversible. In other words, even though we assume that stages and phases follow each other in temporal order, no assumptions need be made concerning the direction of change. Decay is always a possibility.[7]

Before describing how we actually constructed our typology of policy development and assigned cities to each stage or phase, let us diagram in Figure 1 how we envisage the developmental sequence:

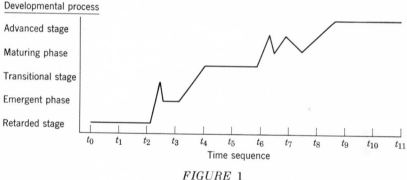

Developmental process

Advanced stage

Maturing phase

Transitional stage

Emergent phase

Retarded stage

t_0 t_1 t_2 t_3 t_4 t_5 t_6 t_7 t_8 t_9 t_{10} t_{11}

Time sequence

FIGURE 1

In the three stages—retarded, transitional, and advanced—development is relatively stable, while in the two phases—emergent and maturing—development is unstable, as symbolized by the zigzag lines. The purpose of this typology is to locate each city's policy profile in a particular stage or phase.

Assignment of Policy Profiles in Developmental Typology. If a policy profile fell below the medians for both planning and amenities expenditures (cell I of Figure 2a), we assigned the city to the *retarded* stage; if it fell above the median lines on either planning or amenities expenditures (cells II or III), we designated the city as being in the *transitional* stage;

[7] See Samuel P. Huntington, "Political Development and Political Decay," *World Politics,* **17**, 386–430 (1965).

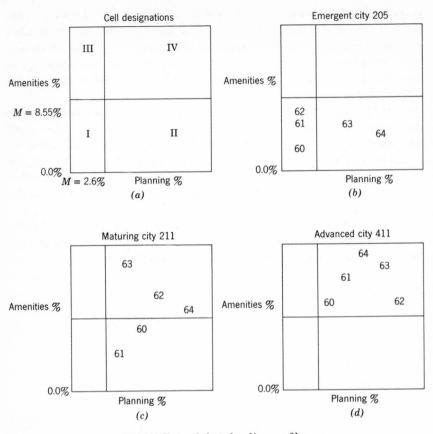

FIGURE 2   *Selected policy profiles.*

and if it fell above both median lines (cell IV), we considered it as being in the *advanced* stage of policy development.

We similarly assigned cities to our two phase constructs if they were moving across the median lines in the five-year period. In Figure 2*b*, for instance, city 205 moved after three years from cell I to cell II; and we therefore characterized it as *emergent*. In Figure 2*c*, city 211 moved, after two years, from cell II to cell IV; the city was typed as *maturing*.

In the assignment we had to take a few liberties with deviant cases. As Table 1 shows, for the 88 cities over a period of five years there were 337 opportunities for change in outcomes.[8] Of these opportunities

---

[8] This calculation is made as follows. Any one city's policy outcomes could change four times, from 1960 to 1961, 1961 to 1962, 1962 to 1963, 1963 to 1964. For

46, or 14 per cent, represented reversals from one year to the next. In the other 86 per cent of opportunities, there either was no change, that is, the city's policy profile remained stable, or change occurred as hypothesized ("progressive"). As Table 1 also shows, because of the assignment procedure, reversals are expected more frequently in the phases than in the stages (21 and 25 per cent versus 9, 11, and 10 per cent, respectively). Reversals in the stable stage cities are due, of course, to our assigning some "impure" cases to stages rather than phases where we felt that the reversals were only temporary deviations from the regular stage pattern.

TABLE 1   DEVELOPMENTAL TYPOLOGY OF CITY POLICY PROFILES AND OPPORTUNITIES FOR CHANGE

Developmental Type	Number of Cities	Number of Opportunities	Per Cent Reversals
Retarded	16	53	9
Emergent	12	47	21
Transitional	31	121	11
Maturing	11	44	25
Advanced	18	72	10
Total	88	337	14

To reduce arbitrary assignment as much as possible, the technique used was in some respects similar to unidimensional scaling methods. We gave each city a score for each policy outcome—1 if the outcome located the city in cell I; 2 for cell II and III outcomes; and 3 for being in cell IV. The individual scores were then totaled and assignment was made. For instance, a city that in all five years had policy outcomes in cell I would have a score of 5; cell II and III outcomes would total 10; and cell IV outcomes 15. Cities with total scores lying between the limits imposed by each cell were assigned to stages or phases in such a way that "reversal errors" in the stages would be minimized. For instance, city 422, with individual scores of 2, 2, 1, 1, 1 or a total score of 7 and 1 reversal error, was assigned to the retarded stage. City 410, with the same total score but 2 reversal errors because of an individual score pattern of 1, 2, 1, 2, 1, was assigned to the emergent phase.

---

88 cities this would make for 88 × 4 or 352 opportunities for change. However, as a number of cities were incorporated after 1960, actual opportunities were only 337.

Resource Capability

The resources available to a city government are an important constraint on the expenditures it can undertake. The cost of government services must be met in each city by some combination of current revenues and borrowed funds, and past borrowing must be paid off according to a rigid schedule. Each city has a broad array of revenue sources available to it, but each source can be expected to produce only a certain amount of revenue and, consequently, the total amount of revenue has definite limits. We refer to the maximum amount of income a city may expect yearly, when the city makes an effort to tap all income sources open to it, as its resource capability.

Resource capability is largely an objective factor determined by the level of actual wealth in the city, tax rebates from county and state governments, federal or state grants, and so on. But it is also subjective in that city residents will vary in how high a tax rate they are willing to tolerate, and city councils will vary in how much they feel they can count on rebates and grants. Resource capability is a limiting factor in the city environment, but its limiting effect must be interpreted by citizens and officials before it becomes a factor in the policy-making process. High resource capability is necessary for a relatively high level of city expenditures, but it is not sufficient.

We could not, of course, use the readily available city income figures as a measure of resource capability because this would directly contradict our assumption that some cities are pressed for revenue and others are not. Ideally, we should estimate for each city the revenue it might expect from each of its income sources if it applied for all possible grants, raised its taxes as high as possible, founded income-earning public utilities, and so on. But this is beyond the scope of our research. Similarly, the customary indicators for resource capability used in the study of developing nations, such as per capita gross national product, cannot be used at the city level because a high proportion of the production of any city crosses city boundaries and is not available to support local government expenditures.[9] What is needed is a measure of the wealth remaining wholly within city boundaries and subject to local taxation or state taxes refundable to the city.

The impossibility of forming any absolute measure of potentially available resources led us to devise the best indicator of relative resource capability that we could. The measure we are using is the total assessed valuation per capita subject to local taxation for fiscal 1965–1966, as

[9] For a discussion of system capabilities, see Gabriel A. Almond, "A Developmental Approach to Political Systems," *World Politics,* **17,** 195–203 (1965).

determined by the California State Board of Equalization. Assessed valuation includes private houses and property, commercial property, and industrial property. From private property a city government will get personal property tax revenues and a portion of state income tax revenues; from commercial property it will receive property tax revenues and sales tax revenues; and from industrial property it will get property tax revenues. Our assumption in using per capita assessed valuation is that wealth in any of those three forms will be a potential source of revenue, and that total valuation per capita will thus be a rough indicator of a city's resource capability. A city will feel constrained if it has a low level of assessed valuation per capita but will feel freer to institute new programs or expand old ones if it has a high valuation level.

For the purpose of analysis, we have divided the 88 cities into two groups, using the median of per capita assessed valuation for all cities as the dividing line between high and low resource capability.

Group Life: Vitality and Diversity

We are interested in characterizing the "group life" of a city, because it can be an important mediator between environmental conditions and the activities of city government. Demands for city programs frequently arise explicitly within community groups rather than being the inventions of farsighted city councils or city managers. When there is a policy disagreement on the council, an active group or combination of groups may be able to mobilize public support and bring pressure to bear on reluctant council members. Or, as on other levels of government, cooperative patterns may exist. Thus we might find Chambers of Commerce actively cooperating with city governments on renewal projects; homeowner and neighborhood groups being concerned over police protection, annexation, and industrial growth; or garden clubs urging the city to buy up land for parks and open spaces. Reform groups and protest groups may carry on extended warfare with city councils not constituted to their liking. In all these cases the community group is bridging the gap between the citizens and the city council by expressing policy demands, by providing information, or by mobilizing support.[10]

---

[10] We define groups broadly, similar to the term "game" developed by Long and Smith. See Norton E. Long, "The Local Community as an Ecology of Games," *American Journal of Sociology,* **64,** 251–61 (1958); and Paul A. Smith, "The Games of Community Politics," *Midwest Journal of Political Science,* **9,** 37–60 (1965). Each kind of group is playing its own game with goals it defines and with its own limited audience. We differ from Long's formulation of community games, however, in assuming that many of the important games *cannot* be played in isolation from each other or from city government. Changes in environmental conditions give rise to new needs for some city residents, and new groups may often be formed to meet these needs for their members and clients. The best

Of the many possible characteristics of a community's group life, two in particular are of importance—the *vitality* of the individual groups and the *diversity* of types of groups. If a group is sensitive to the benefits it can derive from governmental action, the more active the members are in their common pursuits, the more likely it is that the group will have a significant impact on public policies. The greater the number of types of groups, the more likely it will be that a wide variety of government programs will be sought by citizens and provided by city government. A community with a diverse and active group life offers more channels for direct citizen influence over policy outcomes than a community where groups are few in number and member participation is low.

The nature of a city's group life can be measured in several ways. For the purpose of this preliminary analysis and because of ease of coding we devised a measure from closed questionnaire responses, using the councilmen as informants about their own cities. The councilmen were asked the following question: "We would appreciate your helping us rate the *participation of their members* in the activities of some organizations in your city." We listed nine general types of organizations—from chambers of commerce and trade unions to library associations and neighborhood clubs. For each type the respondent could check "high," "moderate," "low," or not applicable." High and moderate checkings were interpreted as indicating that the type of organization was significantly active and low and inapplicable checkings that it was not active in a significant way. By counting the number of types of organization rated as either high or moderate in member activity, we formed a *composite* measure of the vitality and diversity of the community's group life ranging from zero to nine. For the purpose of analysis we have trichotomized this measure between two and three groups and between five and six groups rather than retaining the full range offered by the data.

Policy Orientations: Commitment to Development and Attitude toward Activity Scope

Policy orientations are the normative dispositions to action which decision-makers bring into the policy-making process. In an analysis of policy

choice of strategies for a group is dictated by the kind of goal it pursues and by the number of other groups which might have conflicting goals or be competing with it for limited government funds. Thus every group seeking a goal which would require expenditure of public funds must work through the city council if it is to succeed, and any groups seriously in conflict with each other must usually go to the council for authoritative decisions. We should also repeat here that in focusing on the city council as the major governing body in a city, we are necessarily excluding conflicts and demands relating to educational or welfare programs from our analysis of group activities.

development geared to the functional problems of controlling the environment or adjusting to it, orientations at least somewhat related to these two aspects of policy are of particular interest. The extent to which orientations are conducive to control or adaptation may in part explain why responses to environmental challenges take the form they do.

In the analysis we shall employ an attitude scale capturing an orientation which favors the use of resources for expanded services, planning, and growth and another scale which measures attitudes toward the scope of governmental activities beyond the traditional services.

The *commitment to development* scale consists of four items with which councilmen could agree or disagree.[11] The scale items ordered as follows:

a.   Cities should expand their services just as states are doing.
b.   A city should not hesitate to increase its debts to finance public works projects if they cannot otherwise be paid for.
c.   Every city should provide for economic growth by attracting new industries and stimulating local business.
d.   A master plan and a full-time professional planning staff are necessary to guide city development.

The *scope of activity* scale seeks to order councilmen in terms of their attitudes toward the range or scope of things that the city government should attend to, that is, the substantive areas in which the life of the city can be adapted to the environment.[12] This scale also consists of four items which ordered as follows:

a.   Many formerly private concerns such as health services and hospitals should now have more active financial support from city governments.
b.   Where possible, a city should set aside land for large-scale, tract-type residential building.

[11] These items were scaled by Guttman's criteria for unidimensionality using Ford's method, see Robert N. Ford, "A Rapid Scoring Procedure for Scaling Attitude Questions," in M. W. Riley, J. W. Riley, Jr., and J. Toby, *Sociological Studies in Scale Analysis* (New Brunswick: Rutgers University Press, 1954), pp. 273–305. The scale yielded a coefficient of reproducibility of .93, well above Guttman's standard. Errors of the individual items were 1, 12, 9, and 7%, respectively, also well within the permissible error range of 15% for this type of scale. Altogether 75% of the 282 councilmen whose responses could be scaled were perfect scale types. The imperfect response patterns were assigned to the perfect scale types in such a way that errors would be reduced. See Andrew F. Henry, "A Method for Classifying Non-Scale Response Patterns in a Guttman Scale," *Public Opinion Quarterly*, **16**, 94–106 (1952).
[12] These items were scaled by the same methods and yielded a very satisfactory coefficient of reproducibility of .96. Errors of the individual items were 5, 4, 5, and 2%, respectively. The pure scale types included 84% of the 282 councilmen who could be scaled.

c.   Setting aside areas for new shopping centers promotes continued city prosperity.

d.   Every city should provide in its budget for amenities such as parks and libraries for its citizens.

The two scales measure different aspects of a councilman's policy orientation. The two kinds of orientation are positively, if only slightly, interrelated, as Table 2 shows.

TABLE 2   RELATIONSHIP BETWEEN COMMITMENT TO DEVELOPMENT AND ATTITUDE TOWARD ACTIVITY SCOPE

Attitude toward Scope	Commitment to Development (Per Cent)			
	High	Medium	Low/None	$N =$
Favorable	46	45	23	104
Supportive	26	32	21	72
Unfavorable	28	23	56	96
	100	100	100	
$N =$	103	85	84	

THE MODEL

Because we are concerned with only two characteristics of policy outcome, our model is necessarily a partial one. In its general form the model predicts city policy development as a response to certain external and internal features of the urban environment. The external features include (but are certainly not exhausted by) city growth and city size as symptoms of common challenges from the environment as well as city resource capability as an environmental constraint on policy outcomes. The internal features include (but, again, are not exhausted by) the city's organized group life as indirect evidence of demands made by groups for certain policies as well as the policy orientations which policy-makers themselves bring to the policy-making situation. The task of the model is to order those component variables and to link them to each other in a theoretically viable manner.

For the purpose of ordering the variables, the model postulates that city growth, size, and resource capability are antecedent or independent variables; that the city's group life and policy-makers' orientations to action are intervening variables; and that policy outcomes and resultant stages or phases of policy development are consequent or dependent variables. In empirical reality, of course, neither growth, size, nor resource capability are truly independent precisely because public policy can affect them, by controlling growth and size or by increasing resource capability. But, for the short-term analysis undertaken here, we can assume these variables to be antecedent. Moreover, growth, size, and resource capability may, but need not, be highly interrelated.

The city's group life is likely to be related to its size. The larger a city's population, the more diverse groups are likely to be formed (and, moreover, the more diverse demands for policies coping with problems arising out of size and growth are likely to be made). Decision-makers' policy orientations, on the other hand, should be independent of environmental variables; that is, attitudes favorable or unfavorable to urban development or increased scope of urban government activities are likely to be randomly distributed across cities of any size, growth, or resource capability.[13]

Policy outcomes follow each other in a characteristic sequence that constitutes a stage or phase of a city's policy development. These outcomes are responses to environmental challenges, such as those brought about by large size or a high growth rate, and are indicative of a city's willingness to draw on its resources. Changes in city growth, size, and resource capability bring about changes in policy outcomes that move the city along from stage to phase to stage in the developmental process. However, though the model assumes that, in general, the developmental process moves the city from a retarded stage into a phase of emergence, from emergence into a more stable transitional stage, from there into a maturing phase, and finally into an advanced stage, the process of policy development is not inevitably unidirectional. Cities in the two phases of development are especially subject to temporary reversals.

Environmental challenges must be perceived by policy-makers before they affect the policy process. Policy-makers' reactions are in turn influenced by the city's group life (and corresponding demands) and by their own orientations as to what course of policy should be followed. As a result, the stage or phase in which a city is located at a particular moment in time, while not independent of environmental conditions, is also mediated by the city's group life and by policy-makers' orientations to action. Whether or not resource capabilities are mobilized for development depends exclusively on the demands made on government and the policy

---

[13] For representational purposes the policy orientations of city councilmen might well be related to the policy preferences held by citizens and articulated by organized community groups, but we have not pursued that possibility in this paper.

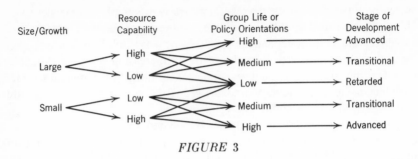

FIGURE 3

preferences of the policy-makers. We can diagram the model as shown in Figure 3.

We shall not in this paper test all the propositions that are implicit in the model. We do not as yet have data that link perceptions of environmental challenges to policy-makers' orientations and their attitudes toward group demands (though we will have these data for later analyses). Nor are we deriving at this time any propositions about the relationship between a city's group life as a structural variable of the political system and policy-makers' orientations to action. However, we shall be testing the following hypotheses.

1. The larger a city's size, the more developed it is likely to be.
2. The greater a city's growth, the more developed it is likely to be.
3. The higher a city's resource capability, the more developed it is likely to be.
4. The larger a city's size, the more vital and diverse its group life is likely to be.
5. Regardless of size, the more vital and diverse a city's group life, the more developed it is likely to be.
6. Policy orientations of policy-makers concerning development and scope of government activities are not related to city size, growth, or resource capability.
7. The more favorable policy-makers' orientations toward development, the more developed the city is likely to be.
   a. Regardless of size, growth, or resource capability, the more favorable policy-makers' orientations toward developments, the more developed the city is likely to be.
8. The more favorable policy-makers' orientations toward increasing the scope of government activity, the more developed the city is likely to be.
   a. Regardless of size, growth, or resource capability, the more

favorable policy-makers' orientations toward increasing the scope of government activity, the more developed the city is likely to be.

THE DATA AND TESTS OF HYPOTHESES

The data used in the testing of the hypotheses are the following.

(a) City size and city growth data from 1960 and 1965 census.

(b) City per capita assessed valuation data and expenditure data for planning and amenities from the *Annual Report of Financial Transactions concerning Cities of California,* for the fiscal years 1960–1961 to 1964–1965, published by the State Controller.

(c) Data concerning city group life and policy-makers' orientations from self-administered questionnaires most often left after interviews with city councilmen to be returned by mail, though at times filled out immediately after the interview. These data were collected between January and April, 1966.

When the analysis was prepared not all councilmen in the 88 cities had been interviewed and not all of those interviewed had returned the mail questionnaire. We therefore could work only with 304 questionnaries, or 60 per cent of our population universe (at the time of the analysis, the return rate for distributed questionnaries was 76 per cent). Because not all respondents marked every item, our orientation scales report on 282 councilmen and our group life measure is based on 249 responses.

Interviews with councilmen were sought and held in eight counties of the San Francisco Bay region, including Alameda, Contra Costa, Marin, Napa, San Mateo, Santa Clara, Solano, and Sonoma. Interviews were also conducted in the city-county of San Francisco with members of the Board of Supervisors, but they are not used in this study. Two cities, Portola Valley and Yountville, are not included because no attempts to interview had been made in these recently incorporated cities at the time of the analysis.

Many of the propositions suggested by the model could have been better handled with interview data than with the questionnaire data. But the interview data had not as yet been coded; and we relied, therefore, on the quickly codable closed items on the self-administered questionnaire.

Hypothesis 1:   The larger a city's size, the more developed it is likely to be.

Hypothesis 2:   The greater a city's growth, the more developed it is likely to be.

TABLE 3   RELATIONSHIPS BETWEEN CITY SIZE AND GROWTH
AND POLICY DEVELOPMENT

	Policy Development (Per Cent)				
	Retarded $N = 16$	Emergent $N = 12$	Transitional $N = 31$	Maturing $N = 11$	Advanced $N = 18$
Population size					
<10,000	94	58	39	10	5
10–50,000	6	42	49	54	45
>50,000	0	0	12	36	50
	100	100	100	100	100
Growth rate (%)					
<10	62	41	32	10	28
10–50	38	42	49	45	50
>50	0	17	19	45	22
	100	100	100	100	100

A city's size and growth rate are indicative of challenges from the environment. The two hypotheses state that policy development is positively related to challenges from the environment arising out of size and growth. Cities with larger and fast-growing populations should be more developed than cities with smaller and slow-growing populations. Table 3 shows that the two hypotheses are not falsified by the data. The smaller a city's size, the more likely it is to be in the retarded stage of policy development; the larger it is in size, the more likely it is to be in the advanced stage. Moreover, the data show that the pattern of development is highly linear.

With regard to growth, the pattern is similar, though there is, as we should perhaps expect, a leveling-off of the effect of growth in the advanced stage as this is the terminus of development. But, in general, the lower the growth rate, the more a city tends to be underdeveloped, and the greater the growth rate, the more it tends toward the more-developed end of the continuum.

Hypothesis 3: The higher a city's resource capability, the more developed it is likely to be.

Table 4, Part A, shows that there is no support for the hypothesis. Only in the advanced cities does resource capability seem to play a significant role.

However, the distributions in Part A of Table 4 may be misleading. In contrast to environmental challenges over which policy-makers have relatively little control and with which they may have to come to grips regardless of available resource capabilities, policy development is likely to depend a great deal on policy-makers' willingness to tap their city's resources. Whether they will tap them, in turn, may depend on the intensity of the environmental challenges. Hence, we must control by size and growth rate for a hidden relationship between resource capability and policy development. Table 4, Part B, reports the findings.

With respect to size, in the smaller cities development declines with low capability, just as the hypothesis postulated; but it also declines in the cities of the same size with high capability, counter to the hypothesis! In the larger cities, on the other hand, resource capability remains related to policy development in the advanced stage. In other words, a great deal of the variance in the relationship between resource capability and policy development can be accounted for by the challenging factor of

TABLE 4   RELATIONSHIP BETWEEN CITY RESOURCE CAPABILITY
AND POLICY DEVELOPMENT

	Policy Development (Per Cent)				
	Retarded $N = 16$	Emergent $N = 12$	Transitional $N = 31$	Maturing $N = 11$	Advanced $N = 18$
Part A					
Resource capability					
Low	50	58	58	64	28
High	50	42	42	36	72
	100	100	100	100	100
Part B					
Size <25,000					
Low	50	58	26	18	11
High	50	42	39	36	11
Size >25,000					
Low	—	—	32	46	17
High	—	—	3	—	61
	100	100	100	100	100
Growth <10%					
Low	19	17	16	10	4
High	44	25	16	—	22
Growth >10%					
Low	31	42	39	55	22
High	6	17	29	35	50
	100	100	100	100	100

size alone. Evidently, in smaller cities with high resource capability, the latter is not enough to stimulate policy-makers to put their city on the road toward a more developed stage of public policy.

Similarly, if resource capability is controlled by growth, in the fast-growing cities with high capability and the slow-growing cities with low capability the developmental process follows the hypothesized pattern; but in the slow-growth, high capability and the high-growth, low capability cities policy development does not proceed as anticipated.

The negative findings of Part B in Table 4 confirm our hunch that how policy-makers respond to environmental pressures depends less on available resources as such than on their willingness to tap these resources, that is, on their orientations toward development and the scope of government activity.

Hypothesis 4: The larger a city's size, the more vital and diverse is its group life likely to be.

This hypothesis is of interest only because the relationship between a city's group life and policy development, specified in Hypothesis 5, may be powerfully dependent on city size. It is, therefore, necessarily preliminary to appraising the relationship between group life and policy development. As Table 5 shows, the larger cities are characterized by

TABLE 5 RELATIONSHIP BETWEEN CITY SIZE
AND GROUP LIFE

Number of Active Groups, by Member Participation, Reported by Councilmen	City Size	
	>25,000 N = 104	<25,000 N = 145
0–2	8%	51%
3–5	58	42
6–9	34	7
	100%	100%

a more active and diverse group life than the smaller cities. In interpreting the table, it should be recalled that our measure of the active types of groups reported by councilmen is a composite one that includes both aspects of group life. It should also be noted that in this and all succeeding tables percentages are based on the number of individual respondents rather than cities.

Hypothesis 5: Regardless of size, the more vital and diverse a city's group life, the more developed it is likely to be.

Underlying this hypothesis was the assumption made in the model that the group life of a city is an intervening variable between environmental challenges of size and growth and the policies followed by city government, serving as a channel toward the city council for policy demands and pressures for policy changes and as a means to mobilize support for or opposition to contemplated policies.

If size is discounted, Table 6 suggests that there is a direct relationship between a city's group life, as reported by councilmen, and various stages and phases of development. But since in testing Hypothesis 4 a strong relationship between city size and group life was found, size must be controlled to determine whether group life is still likely to have an effect on policy development within categories of cities of the same general size. From the logic of the model we would expect group life to be more closely associated with policy development than city size would be, because an increase in size logically precedes an increase in the diversity of group life, and a growing city need not necessarily have developed as diverse a group life as would be expected from its size.

If size is controlled, as shown in Table 6, the relationship between

TABLE 6   RELATIONSHIP BETWEEN CITY GROUP LIFE
AND POLICY DEVELOPMENT

	Policy Development. (Per Cent)				
	Retarded	Emergent	Transitional	Maturing	Advanced
Number of active groups reported	$N = 41$	$N = 23$	$N = 83$	$N = 34$	$N = 68$
0–2	76	57	24	15	19
3–5	19	43	60	59	50
6–9	5	0	16	26	31
	100	100	100	100	100
Size <*25,000*	$N = 41$	$N = 23$	$N = 53$	$N = 10$	$N = 18$
0–2	76	57	38	40	33
3–5	19	43	54	60	45
6–9	5	0	8	0	22
	100	100	100	100	100
Size >*25,000*	$N = 0$	$N = 0$	$N = 30$	$N = 24$	$N = 50$
0–2	—	—	0	4	14
3–5	—	—	70	59	52
6–9	—	—	30	37	34
	—	—	100	100	100

group life and policy development is destroyed in the larger cities, but it remains essentially unimpaired in the smaller cities. That is, city size and group life are neither totally related nor totally independent in their effect on policy development. In smaller cities the presence of a number of diverse community groups with an active membership still appears to be of importance for policy development, while in larger cities other factors are also involved in changes of policy outcomes that make for development. We might speculate that the sheer size of the government establishment in larger cities tends to push expenditures to the point where the city would be classified as more developed in the development sequence. It should be noted, however, that none of the cities with a population of over 25,000 was found in the retarded stage or emergent phase and that the larger cities had a more diverse group life than the smaller cities *at the same stage or phase* of development, where comparison was possible.

Hypothesis 6:   Policy orientations of policy-makers concerning development and scope of government activities are not related to city size, growth, or resource capability.

The only theoretical significance to be attached to this hypothesis is that normative orientations to action are truly independent variables that policy-makers bring into the policy-making situation. As Table 7 shows, councilmen's policy orientations are (with two exceptions) quite unrelated to their city's environmental conditions and resource capabilities. One exception relates to city size and commitment where there is a tendency for a more favorable attitude toward development in the larger cities. Similarly, there are more councilmen favorable to increasing the scope of government in the fast-growing than in the slow-growing cities. With respect to these relationships, then, the hypothesis is falsified. But, more important, neither type of policy orientation is significantly related to resource capability. This independence is critical for the validity of the analysis that seeks to relate policy orientations to policy development.

Hypothesis 7:   The more favorable the policy-makers' orientation toward development, the more developed the city is likely to be.

Hypothesis 7a: Regardless of size, growth, or resource capability, the more favorable the policy-makers' orientations toward development, the more developed the city is likely to be.

The validity of Hypothesis 7a is critical for the validity of Hypothesis 7. Only if the relationship between commitment to development and policy development is not dependent on city size, growth, and resource capability, can we say that the policy-makers' orientations to action are significant factors in the policy-making process.

Table 8 shows that, without controlling for size, growth, or resource

TABLE 7   RELATIONSHIPS BETWEEN CITY SIZE, GROWTH, AND RESOURCE
CAPABILITY, AND COUNCILMEN'S POLICY ORIENTATIONS (PER CENT)

	Size		Growth		Capability	
	<25,000	>25,000	<10%	>10%	Low	High
Commitment to						
development	$N = 171$	$N = 111$	$N = 101$	$N = 181$	$N = 152$	$N = 130$
High	29	51	35	40	37	38
Medium	32	30	28	33	35	27
Low and none	39	19	37	27	28	35
	100	100	100	100	100	100
Attitude toward						
activity scope	$N = 171$	$N = 111$	$N = 97$	$N = 185$	$N = 148$	$N = 134$
Favorable	36	41	29	43	38	37
Supportive	22	31	22	28	31	20
Unfavorable	42	28	49	29	31	43
	100	100	100	100	100	100

capability, there is a significant relationship between the councilmen's orientations committing them to development and the stage or phase of the developmental process in which their city happens to be. As development proceeds from the retarded to the advanced stage, the proportions of highly committed councilmen increase in accordance with Hypothesis 7. Though the distributions in the emergent and maturing phases are difficult to locate in the overall table pattern, in the three stable stages—retarded, transitional, and advanced the hypothesized relationship is well marked.

But we can have confidence in the relationship between a prodevelopment policy orientation and policy development only if, as Hypothesis 7a requires, it is not disturbed by the environmental conditions of size and growth, which are also related to policy development. As Part B of Table 8 shows, the basic pattern remains even with the controls introduced. Regardless of size or growth rate, high commitment to development follows the developmental sequence from retarded to advanced stages.

More critical for a test of Hypothesis 7 is that the expected relationship not be an artifact of a city's resource capability.

As Part C of Table 8 demonstrates, the pattern of councilmen's commitment to development in the various stages and phases of policy development remains undisturbed. In fact, in all but the retarded stage highly committed councilmen appear somewhat more frequently in the low capa-

TABLE 8 RELATIONSHIP BETWEEN COMMITMENT TO DEVELOPMENT AND POLICY DEVELOPMENT, WITH CONTROLS FOR SIZE, GROWTH, AND RESOURCE CAPABILITY

	Policy Development (Per Cent)				
	Retarded	Emergent	Transitional	Maturing	Advanced
Part A					
Commitment					
to development	$N = 53$	$N = 20$	$N = 97$	$N = 35$	$N = 78$
High	15	25	36	43	57
Medium	23	60	37	20	27
Low and none	62	15	27	37	16
	100	100	100	100	100
Part B					
Size <25,000	$N = 52$	$N = 20$	$N = 64$	$N = 14$	$N = 21$
High	15	25	31	29	62
Medium	23	60	38	21	19
Low and none	62	15	31	50	19
	100	100	100	100	100
Size >25,000	$N = 0$	$N = 0$	$N = 33$	$N = 21$	$N = 57$
High	—	—	46	52	55
Medium	—	—	36	19	30
Low and none	—	—	18	29	15
	—	—	100	100	100
Growth <10%	$N = 32$	$N = 4$	$N = 37$	$N = 4$	$N = 24$
High	19	0	30	a	71
Medium	16	75	38	0	25
Low and none	65	a	32	75	a
	100	100	100	100	100
Growth >10%	$N = 20$	$N = 16$	$N = 60$	$N = 31$	$N = 54$
High	a	31	40	45	50
Medium	35	56	37	23	28
Low and none	55	a	23	32	22
	100	100	100	100	100
Part C					
Low Capability	$N = 30$	$N = 14$	$N = 57$	$N = 27$	$N = 24$
High	13	21	39	44	67
Medium	33	58	42	19	25
Low and none	54	21	19	37	a
	100	100	100	100	100

TABLE 8 (Continued)

	Policy Development (Per Cent)				
	Retarded	Emergent	Transitional	Maturing	Advanced
Part C (Continued)					
High Capability	N = 22	N = 6	N = 40	N = 8	N = 54
High	18	a	33	38	52
Medium	a	67	30	a	28
Low and none	73	0	36	38	20
	100	100	100	100	100

a Less than three cases in cell.

bility than in the high capability cities, further evidence that Hypothesis 7 has not been falsified by the data.

Hypothesis 8:   The more favorable the policy-makers' orientations toward increasing the scope of government activity, the more developed is the city likely to be.

Hypothesis 8a: Regardless of size, growth, or resource capability, the more favorable the policy-makers' orientations toward increasing the scope of government activity, the more developed is the city likely to be.

On testing Hypotheses 8 and 8a, we find that they are totally falsified by the data. As Part A of Table 9 shows, councilmen with favorable attitudes toward expanding the scope of city government predominate in cities that are in the emergent and maturing phases of development, while in the more stable-stage cities their attitudes are less favorable, and in the retarded stage a majority is even unfavorable. How can we explain this finding? Is it merely due to arbitrary chance, or is it an unexpected finding of theoretical significance?

In the first place, we may seek an explanation in the nature of the emergent and maturing phases of policy development as we have defined them. Cities in these phases seem to undergo a sudden burst of activity, reflected in policy outcomes, that moves them from one stage into another. By way of contrast, in the more stable stage cities the direction of policy is well set, at least for the time being. Allocations are made to established programs, but with regard to new ones there is, by definition again, little change. In these cities, budget expenditures are largely treated as if they were nonprogrammatic.

TABLE 9   RELATIONSHIP BETWEEN ATTITUDE TOWARD SCOPE OF
GOVERNMENT ACTIVITY AND POLICY DEVELOPMENT, WITH CONTROLS
FOR SIZE, GROWTH, AND RESOURCE CAPABILITY

	Policy Development (Per Cent)				
	Retarded	Emergent	Transitional	Maturing	Advanced
Part A					
Attitude toward scope of activity	$N = 51$	$N = 20$	$N = 99$	$N = 35$	$N = 77$
Favorable	29	65	36	60	29
Supportive	14	15	32	26	29
Unfavorable	57	20	31	14	42
	100	100	100	100	100
Part B					
Size <25,000	$N = 51$	$N = 20$	$N = 64$	$N = 14$	$N = 22$
Favorable	29	65	42	50	0
Supportive	14	15	25	29	36
Unfavorable	57	20	33	21	64
	100	100	100	100	100
Size >25,000	$N = 0$	$N = 0$	$N = 35$	$N = 21$	$N = 55$
Favorable	—	—	26	67	40
Supportive	—	—	46	24	25
Unfavorable	—	—	28	9	35
	—	—	100	100	100
Growth <10%	$N = 29$	$N = 5$	$N = 36$	$N = 4$	$N = 23$
Favorable	21	60	36	a	22
Supportive	17	0	33	a	a
Unfavorable	62	a	31	—	69
	100	100	100	100	100
Growth >10%	$N = 22$	$N = 15$	$N = 63$	$N = 31$	$N = 54$
Favorable	41	66	37	64	31
Supportive	a	20	32	22	37
Unfavorable	50	a	32	13	32
	100	100	100	100	100

TABLE 9 (Continued)

	Policy Development (Per Cent)				
	Retarded	Emergent	Transitional	Maturing	Advanced
Part C					
Low Capability	$N = 29$	$N = 13$	$N = 57$	$N = 27$	$N = 22$
Favorable	41	61	33	59	a
Supportive	17	a	42	26	41
Unfavorable	42	31	25	15	50
	100	100	100	100	100
High Capability	$N = 22$	$N = 7$	$N = 42$	$N = 8$	$N = 55$
Favorable	14	72	41	63	37
Supportive	a	a	19	a	24
Unfavorable	77	0	40	a	39
	100	100	100	100	100

a Less than three cases in cell.

As Wildavsky writes, referring to Congress, nonprogrammatic

. . . does not mean that appropriation committee people do not care about programs; they do. Nor does it mean they do not fight for or against some programs; they do. What it does mean is that they view most of their work as marginal, monetary adjustments to existing programs so that the question of the ultimate desirability of most programs arises only once in a while.[14]

The fact, therefore, that councilmen with favorable attitudes toward the scope of government predominate in those cities where the developmental process is accelerated should not come as a surprise. In these cities, it seems, there is high agreement among councilmen to promote their city by expanding the range of government services and, in this way, to adjust their city to the changing environment.

While the data, then, falsify the original hypothesis, they suggest changes that must be made in the model to bring it closer to reality. The model must provide for the possibility that policy-making with regard to certain functions of local government is fundamentally different in the stages and phases of the developmental process.

But, secondly, we can have confidence in this new hypothesis only

[14] Aaron Wildavsky, *The Politics of the Budgetary Process* (Boston: Little, Brown, and Company, 1964), p. 60.

if the initial finding is not spurious. As Part B of Table 9 shows, when attitudes toward the scope of activity are controlled by size and growth, the pattern remains the same. Those favorable to increasing the scope continue to predominate in the emergent and maturing phases of policy development.

Even more important, we must be sure that a favorable attitude toward an increased scope is not simply due to an abundance of resources for expansion. We therefore control for resource capability. As Part C of Table 9 indicates, favorable attitudes are not due to high resource capability. Regardless of whether a city has or does not have resources for potential expansion of its scope of operations, councilmen most favorable to an increased scope are more frequent in the emergent and maturing phases than in the stable stages of policy development. However, in contrast to commitment to development, and with the exception of retarded stage cities, there is a tendency for a favorable attitude toward scope to be more prominent in the high rather than in the low capability cities. Evidently, resource availability is an additional incentive for policy-makers favorably disposed to expand governmental activities to do so.

SUMMARY

The analysis attempted here—relating census data, expenditure data, and questionnaire data—is extraordinarily difficult methodologically. The theoretically postulated relationships are equally difficult to demonstrate. We would not have been surprised if our findings had turned out to be negative. The moderate success that we do have is encouraging.

A city's position in one of the five development types—retarded stage, emergent phase, transitional stage, maturing phase, and advanced stage—is positively related to its size, larger cities being more developed, and to its growth rate, faster-growing cities being more developed, except that the most-developed cities show a falling off in growth rate. Development is not related to a city's resource capability, even when city size and growth are controlled. The combined vitality-diversity measure of group life is positively related to city size, and more advanced development is also related to a more diverse and vital group life when city size is controlled. The policy orientations of city councilmen are essentially unrelated to city size, growth rate, and resource capability. Councilmen more favorable to development are found disproportionately in more-developed cities, even when size, growth, and resource capability are controlled. Councilmen favoring a wide scope of governmental activity are found in the intermediate phases of development rather than, as hypothesized, in the more stable stages, even when size, growth, and resource capability are controlled.

The data are interpreted as supporting the general validity of a model of city policy development as a response to challenges from a changing city environment. Analytically, the adaptation and control functions of policy are responses to city growth, problems arising out of city size, and resource capability, with the city group life and the councilmen's policy preferences being major intervening variables. Theoretically, policy is the result of the forcing effects of population size and growth, as mediated by the city group life, and the goals sought by policy-makers, as expressed in their commitment to development and their attitude toward the scope of government activity. In some cases resource capability may be an important constraint on policy development, but the willingness of policy-makers to tap available resources seems to be a more important variable in explaining the course of policy development. Whether the relative importance of the several variables we have studied holds true at other levels of government can only be determined by further research. We hope the relationships we have been able to demonstrate will help to restore the political scientists' belief in the importance of politicians in the policy process and perhaps stimulate studies tapping policy orientations in a more sophisticated manner than we have been able to do in this preliminary analysis.

PART II

Politics and Public Finance

ROBERT C. WOOD

The Local Government
Response to the
Urban Economy

Our line of departure for describing the political response to economic change lies with the internal politics of the New York Region's localities. More properly it lies with one kind of local politics, the kind that has to do with money matters—how budgets are approved, how tax rates are set, and when decisions to issue bonds for new schools are made. It is here that the political tampering with economic circumstances takes place, a process which pits different factions of a locality's constituency against one another. In the fixing of assessments, the management of public debt, the authorization of new revenue sources, and the establishment of new units, governments make their peace with urbanization—and maintain simultaneously some manner of political stability.

It is the distinction between the environment as it is and the environment as the governments choose to define it which we wish to emphasize. For the public sector is a peculiar sort of artifact, capable of subtle definition and redefinition. And depending how its managers choose to view it, different consequences fall upon the "real" world of households, industry, and commerce.

Reprinted from Robert C. Wood, *1400 Governments* (Cambridge: Harvard University Press, 1961), pp. 67–83, 93–104, 110–113, with the permission of the copyright holder, the Regional Plan Association, Inc., of New York.

STRATEGIES OF ADJUSTMENT

Viewing Economics through Political-Colored Glasses

Perhaps the most obvious possibility for adjustment is found in the legally established local revenue system, in particular the method by which real property taxes are levied. Throughout the New York Region, the amount of the property tax is a function of a tax *rate,* set by the town fathers and subject to state limitation, and an assessed *valuation,* typically established by experts in the local bureaucracies. Since the assessed valuations are generally set considerably below full market value, the revenues resulting from any given rate are usually less than those which a municipality could extract if it chose to raise the assessments. Hence, the political elite in any locality has the continuing option to expand the tax revenue at a given moment, almost regardless of economic developments. Alternatively, it can reduce the size of its yield by a local ordinance or bureaucratic decision. Finally, it may squeeze through loopholes of state law to impose differential burdens on different types of property—and thereby shift the incidence of taxation.

Manipulating the assessments provides the most obvious opportunity for controlling the supply of revenue and thus reshaping the environmental pressures. To the traditional "taxpayers association" whose members pay the largest share of a municipality's bill and are convinced that public programs provide them few direct benefits, this manipulating has logical appeal. Regardless of motive, it is administratively difficult to keep assessed valuations in a constant proportion to market values, especially in times of rapid economic growth. Whether through conscious policy or administrative lag, a wide disparity between assessed and market value acts as an ultimate deterrent to increasing expenditures. Tax rates climb, of course, but sooner or later they bump against ceilings imposed either legally by the state or politically by local indignation. Thus a municipality can exhaust its legal revenue capacities without any increases whatever in the tax burden relative to the market value of the property. It can also generate psychological panic among residents untutored in municipal finance. These good citizens, watching the rates spiral and not realizing that the assessments lag behind the market value, may believe that taxes are climbing astronomically, and may organize crusades against "excessive spending."

We have considerable evidence that there has indeed been a lag in assessed valuations. Taken together, the Region's local governments entered the period after World War I with their assessed valuations pegged at considerably less than one-half of market value. Ten years later, despite the new demands for public services, the steady economic growth, and the unparalleled investment in housing and industrial facilities, the gap was even larger.

In New Jersey, the statewide ratio of assessed to market value declined between 1951 and 1957 from 34 per cent to 28 per cent. Decreases of between one percentage point and eight percentage points were registered in the nine New Jersey counties of the Region, as indicated in Table 1. By 1957, in six out of the nine counties, assessed value was less than one-third of full value.[1]

TABLE 1   CHANGE IN AVERAGE MUNICIPAL ASSESSMENT RATIOS, NEW JERSEY METROPOLITAN COUNTIES (ASSESSED VALUATION AS PERCENTAGE OF ESTIMATED MARKET VALUE)

	1951	1957	Change
Bergen	26	22	−4
Essex	48	41	−7
Hudson	56	55	−1
Middlesex	24	22	−2
Monmouth	22	19	−3
Morris	22	18	−4
Passaic	43	35	−8
Somerset	20	16	−4
Union	37	30	−7

Source: State of New Jersey, Commission on State Tax Policy, *Ninth Report* (Trenton, 1958).

In New York, the same pattern of disparity between assessed and market value exists, although this time we do not have a direct comparison of valuations over time. We can make much the same point, however, by examining the change in the ratio between total property taxes and estimated full property value. For the 82 New York towns outside New York City there was an average decrease in this ratio between 1945 and 1955 of 30.3 per cent, ranging from 3.7 to 71.9.[2] In only two instances did taxes command a larger share of full value at the end of the decade than they did in the beginning.

[1] State of New Jersey, Commission on State Tax Policy, *Ninth Report* (Trenton, 1958).
[2] This comparison is based on the 1945, 1950, and 1955 issues of *Special Report on Municipal Affairs of the State Comptroller of New York.*

It is not only *total* assessed valuations that are manipulated. The administrative process of assessment also offers opportunities to impose different burdens on different classes of property. New Jersey localities, for example, have long been notorious for "tax-lightning"—a sudden raising of the assessment on a particular item of property. In this respect, New Jersey municipalities have established so noticeable a pattern of assessment discrimination against business firms that the New Jersey Commission on State Tax Policy has officially incorporated "tax-lightning" into the vocabulary of public finance.[3]

Yet it is not just business property which is discriminated against in the Garden State. In its 1953 investigations, the Commission found that, in a sample of 21,275 properties, "low value" business and residential property was assessed at higher than average ratios. Specifically, business and residential properties with market values between $2,000 and $4,000 had assessed valuations averaging 41 per cent of market value. Properties between $8,000 and $10,000 showed an average of 29 per cent.[4] The Commission also found that one-family houses were more lightly assessed than multifamily dwellings or business and industrial properties; that built-up blocks, old neighborhoods, and accommodations in the process of rehabilitation carried higher-than-average assessments; and that the ratio of assessed to market value of undeveloped land was almost twice the ratio for residential buildings. These variations reflect, of course, a general sensitivity of town fathers to the political strengths of different classes of property owners. For our purposes, they also mark a further refinement in distinctions between what constitutes taxable resources and what does not, by identifying the resources "most available" politically for public purposes.

Changes in legal definitions of property valuations are not the only ways in which potential resources can be differentiated from existing resources and tax burdens shifted. A second alternative is to restructure the entire local revenue system, pre-empting more resources by legislative declaration. The search for new types of local revenue has, in fact, been evident in all parts of the Region since World War II, although no single other tax has yet approached the property levy in importance. The changing role of the property tax for the Region's local governments is shown in Table 2. This table covers nonschool revenues during the postwar

[3] State of New Jersey, Commission on State Tax Policy, *Fifth Report* (Trenton, 1950), and *Ninth Report* (Trenton, 1958). See also report of Harry V. Osborne, Jr., Deputy Administrator of Taxation, "Investigations of Assessments, City of Passaic," as printed in the *Herald-News,* Passaic-Clifton, New Jersey, July 11, 1951.
[4] State of New Jersey, Commission on State Tax Policy, *Sixth Report* (Trenton, 1953), chap. V, pp. 103–120.

TABLE 2   PROPERTY VERSUS NONPROPERTY REVENUES[a] OF LOCAL GOVERN-
MENTS,[b] NEW YORK METROPOLITAN REGION, 1945–1955

	1945		1950		1955		
	Millions of Dollars	Per Cent of Total	Millions of Dollars	Per Cent of Total	Millions of Dollars	Per Cent of Total	Per Cent Rise in Revenues, 1945–1955
The Region[c]							
Property	418.3	62.3	520.8	56.2	724.4	54.2	73.1
Nonproperty	253.0	37.7	405.2	43.8	611.3	45.8	141.6
New York City							
Property	299.5	60.2	330.6	49.7	478.9	48.6	59.8
Nonproperty	197.4	39.7	334.1	50.3	506.0	51.4	156.3
New York Inner Counties							
Property	29.2	73.2	38.0	73.2	52.2	69.4	75.1
Nonproperty	10.9	26.8	13.9	26.8	23.0	30.6	111.1
New Jersey Inner Counties							
Property	69.9	67.2	122.0	74.7	147.3	72.1	110.7
Nonproperty	34.0	32.8	41.2	25.3	56.9	27.9	67.4
New York Outer Counties							
Property	9.5	72.5	14.4	75.3	23.0	76.9	142.1
Nonproperty	3.6	27.5	4.7	24.7	6.9	23.1	91.6
New Jersey Outer Counties							
Property	9.6	57.4	15.8	58.3	23.0	55.4	139.5
Nonproperty	7.1	42.6	11.3	41.7	18.5	44.6	160.5

[a] Not including revenues for schools, borrowed funds, or state or federal aid.
[b] Not including school districts.
[c] Not including Fairfield County, Connecticut.
Source: Estimates by New York Metropolitan Region Study.

decade 1945–1955. It emphasizes the dwindling dependence of most jurisdictions on the traditional property tax. As seen in the upper right-hand
corner of the table, property revenue increased only 73 per cent while
other revenue—not including borrowed funds or state or federal aid—was
increasing by 142 per cent. The table also shows, however, that not all
parts of the Region followed the general trend. In the "New Jersey Inner"
and "New York Outer" counties the local governments increased their
property revenue faster than their nonproperty revenue.

Still another way by which local governments can tamper with their
financial arrangements is to persuade the state government to change the
legal limitations on tax rates and borrowings. Alternatively, they can ignore
unwelcome limitations and persuade the state to accept interpretations
which, to say the least, are flexible. In New Jersey, where the limitations
take the form of maximum debt allowances expressed as a proportion
of assessed valuation, this latter course often appears to have prevailed.
In 1956, 236 out of the 585 municipalities in the whole state had debts
exceeding their legal limits. These debts, totaling $360 million against
legal limits amounting to $229 million, attested to the ability of municipali-
ties to break through their ceilings by extensions of credit and other special
arrangements.[5]

In New York State, where limits are applied to the local tax rate,
a more formal amendment procedure must be followed to bring about
adjustments. Until 1949, these limits were calculated relative to assessed
valuations. In that year a constitutional amendment expressed them relative to market value in a jurisdiction, estimated by the state and calculated
on the basis of the last five years' valuations. This change greatly increased
the leeway that communities had in increasing tax rates, but the leeway
was least in jurisdictions where assessments were already closest to market
value. Surveying the revised situation in 1954, the Temporary Commission
on the Fiscal Affairs of State Government reported, "While New York
City has not gained sufficient additional property taxing power to solve
its revenue problem and a few other cities still occupy a marginal situation,
the great majority of the cities and apparently most of the villages have
secured a taxing potential that exceeds their immediate needs—in some
instances, the excess is large."[6] The Commission went on to note that
the expansion of the revenue base had caused concern among local officials
and "more concern among real estate groups."[7]

[5] State of New Jersey, Commission on State Tax Policy, *Ninth Report* (Trenton,
1958), p. 139.
[6] State of New York, *Report of the Temporary Commission on the Fiscal Affairs
of State Government* (Albany, 1955), Vol. II, p. 666.
[7] *Ibid.,* p. 616.

Buttressing the strategies of financial adjustment and manipulation is a fourth device—the outright creation of new governments. Essentially, this amounts to a double-tapping of the same revenue base by simply establishing another governmental layer. The usual form is the special district established to provide a single public service, and the revenue source may either be property taxation or a user charge. Except in New York City, this alternative is applied uniformly in the Region for the provision of schools, and it is also in vogue with respect to sundry other services. Whatever the purpose and however financed, the special district has the effect of establishing another channel to the wealth of a local constituency without either affecting the tax rates of the traditional government or introducing new kinds of taxes.

In the New York portion of the Region there are some 300 school districts, generally within the borders of the towns, occasionally extending across town or even county lines. In the New Jersey portion there are about 240 school districts, for the most part coterminous with municipal boundaries but with a tendency to consolidate on a larger basis.[8] In each state, separate levies on the general tax base are authorized for schools, and special limitations on the tax rate or debt volume are set for school purposes. The expenditures of these districts constitute the largest single activity of the Region's local governments, and the pattern of school financing differs sharply from that of the municipalities of general jurisdiction.

That the patterns differ sharply was evident in the last chapter; differences in environmental characteristics such as industrialization, housing density, and age "explain" a much smaller part of the variations in school expenditures than they do in general expenditures. Underlying this relative insensitivity of school expenditures to environmental factors other than community size is the fact that the separate local school taxes, already big, are powerfully supplemented by state grants, which for the Region as a whole are larger for schools than for all other local purposes combined. Moreover, state aid constitutes 25 per cent of total school revenue compared to only 12 per cent of revenue for other purposes. Thus, outside New York City, the school districts of the Region raised $313 million in 1955 from the local property tax in comparison with $245 million raised by governments of general jurisdiction from the same source. But,

[8] U.S. *1957 Census of Governments,* Vol. I, No. 1, and Vol. VI, Nos. 28 and 30. See also *Special Report on Municipal Affairs of the State Comptroller of New York* for 1945, 1950, and 1955; State of New Jersey, annual reports of the Commissioner of Education for 1954–55, 1955–56, and 1956–57; and Dun and Bradstreet, Municipal Credit Surveys, with special reference to school districts. For a detailed study on New Jersey school districts see Commission on State Tax Policy, *Public School Financing in New Jersey* (Trenton, 1954).

in addition to these funds, the school districts could count on over $90 million in state aid, while the general jurisdictions were getting only $21 million. Even including New York City, 47 per cent of all state aid in the New York and New Jersey portions of the Region went for school purposes. For the New Jersey counties of the Region, the figure was 75 per cent. For the New York counties outside New York City, it was 52 per cent. By contrast, in New York City, where no independent school districts exist, it was only 40 per cent.[9]

But it is not just for the school function that the New York Metropolitan Region has come to turn increasingly to the special district. Particularly on the New York State side of the Region, the special district becomes the principal means by which the suburban governments meet the new needs which arise in the transition from rural to urban communities. In the New York counties, the number of special districts other than school districts increased more than 20 per cent during the first postwar decade. At latest count the entire Region had 289 such units, if we go by the definition of "substantial autonomy" employed by the U.S. Census Bureau. According to the definition applied by the New York State government, Nassau County alone had 259 nonschool special districts in 1956 and all the New York counties combined had more than a thousand.[10]

There is still another dimension of flexibility in the district device. By creating special districts, the governments of more general jurisdictions are able to pinpoint areas of growth and to assure that costs in these areas are borne by their residents alone. In addition to this geographical segregation of responsibility, there is functional segregation, in that the expenses of a single public program can be paid for by the actual users. In some cases, the services provided on a district basis are mundane: sewage disposal, street lighting, fire protection, water supply, and refuse collection. In other cases, the district is a flexible instrument to meet essentially modern needs. Parking authorities, park districts, and housing authorities are obvious examples. However based in law and for whatever object established, the district presents an organizational solution to a financial problem.

Moreover, the special district, it should be noted, often serves as a meeting ground for competing political factions within a locality. For groups prone toward more public spending, the capacity of the districts

---

[9] These figures are from a series constructed by the New York Metropolitan Region Study from reports by the Comptroller of the State of New York and by the New Jersey Department of the Treasury.

[10] For a comparison of the relevant definitions, see U.S. *1957 Census of Governments,* Vol. I, No. 1, and the Temporary State Commission on the Coordination of State Activities, *Staff Report on Public Authorities Under New York State* (Albany, 1956), chap. I.

to develop new lines of access to state coffers and to tax property again and again provides a convenient way to avoid head-on budgetary and reassessment battles. For property owners, the judicious design of special districts offers a way of placing upon the newcomers in the community the burden of the additional costs they either want or require. Indeed, if a district is laid out carefully enough, a further discrimination among types of property can result, this time according to geographical location.

In the last analysis, as ingenious as the four principal strategies of redefining the environment are, there is a limit to their effectiveness. Even if a municipality is dominated by a taxpayers' bloc intent on reliving pastoral days, certain modern expenditures are necessary. Even if the majority in a jurisdiction is devoted to quality services, they cannot spend revenues they do not possess. The tax rates for any given revenue source ultimately have their limits, and the development of new tax sources depends on the existence of some sort of taxable economic activity within a jurisdiction's boundaries. The creation of special districts cannot be extended indefinitely, tapping and retapping the same revenue sources to the point of confiscation.

In several ways we can trace the dwindling amount of "elbow room" left for political maneuver and the redefinition of needs and resources in the most-urbanized sections of the Region. One test is to examine the relation between assessed and market value for particular types of municipalities in the Region. It will be recalled that Table 1 revealed a growing gap between these values in the New Jersey counties, but those county ratios did not show the wide range of differences among municipalities. In 1953 the ratios for municipalities varied from 15 per cent to 41 per cent throughout New Jersey and from 27 per cent to 61 per cent throughout New York State. Where the most advanced stages of urbanization were present, and densities highest, the ratio was highest. Uniformly the New Jersey Commission on State Tax Policy found that large, densely-populated municipalities had a strong tendency to assess at higher than the statewide average.[11] It is in these municipalities, the Commission reported, that tax rates rise and the search for new tax sources intensifies as basic requirements for public programs become more costly and less easily avoidable.

There are not only economic but also political limitations to intramunicipal infighting. Highly discriminatory property assessments invite court intervention. In New Jersey the courts, in 1957, mandated equalized assessment in such clear terms as to bring about comprehensive changes

[11] State of New Jersey, Commission on State Tax Policy, *Sixth Report* (Trenton, 1953), chap. V.

in law and administrative practice for property taxation. Quite apart from the violence done to ethical standards, extreme examples of manipulation can provoke bitter controversy and make uncertain the outcome of political conflict. No local official welcomes a position on the top of a powder keg, and most work strenuously to avoid such situations. In these circumstances, then, as urbanization proceeds the Region's political elites turn more and more to different methods. Rather than redefining the environment, they seek to shape directly the processes of urbanization.

Conditioning the Growth Factors

The politics of guiding growth—zoning, planning, and industrial promotion policies—involve a different time perspective and result in different payoffs from the strategies of fiscal manipulation we have just traced. The use or nonuse of these policies is important, however, often explaining why islands of high-value residential neighborhoods are surrounded by industrial sections, or why enclaves of industrial activity persist in areas apparently economically unsuitable for their location. But there is a sizable time lag, usually measured in years, before the consequences of these types of political action show up in the distribution of population and economic activity. The politician relying on them reconciles himself to less dramatic results, which will not be apparent overnight. This evolutionary quality may account for the fairly rudimentary state of these measures in the Region in comparison to alternative strategies.

By far the most universal of the policies employed to guide growth in the Region is the control of land use; and by far the most popular control device is zoning. In 1956, the Regional Plan Association reported that zoning laws were in effect in 465 municipalities within the Region. Within the five boroughs of New York City and the four counties of Bergen, Essex, Nassau, and Westchester, all land use was at least technically governed by zoning regulations. Only 85 municipalities in the Region had no provision for zoning in that year; and all but a few of these jurisdictions lay in the outer counties. Indeed, 43 of them were in two counties alone: 23 in Orange and 20 in Dutchess.[12] Even such "unprotected" communities had the benefit of county planning boards, and public attention to the public control of growth was increasing. Throughout the Region, zoning is now recognized as a device by which densities can be regulated, land use apportioned between "net-revenue-producing" and "net-revenue-using" property, and an individual municipality's fiscal position brought into tolerable balance.

[12] Regional Plan Association, *Bulletin 86* (1956). This paragraph and those following rely heavily on that bulletin as well as upon conversations with the professional staff of the RPA for the summary of current doctrines about zoning.

But it is not only the geographical extent of zoning that is significant in the Region's development. Less recognized but perhaps more significant for the future are the changes which have recently taken place in the techniques and practice of zoning. Originally, ordinances governing land use consisted of little more than the designation of broad areas as residential, industrial, or commercial districts, and their principal objective was to exclude industrial activity from better residential neighborhoods. Now, more and more of the Region's municipalities use zoning with precision and sophistication.

Thus the residential, industrial, and commercial categories of land use are today often subdivided according to specific types of residential, commercial, and industrial activities. Further, modern planning doctrine goes beyond simply excluding manufacturing and commerce from residential zones. The practice of also excluding residences from manufacturing and commercial zones (the essence of what planners call "noncumulative districting") has increased in popularity. New techniques for classifying land have appeared: for example, "performance ratings," which measure the intensity of noise, smoke, or odor produced by a plant to determine the appropriateness of industrial activity at a given site. Architectural and design standards are now incorporated in many zoning laws, reflecting an increased concern with aesthetics. Governments now frequently exercise the authority to regulate both the time at which new development is permitted and the phases of its construction. In short, zoning as a control measure is now a much more influential instrument in guiding population and economic patterns than it was before World War II.

Coincidentally with the development of zoning, policies of municipal mercantilism—"beggar-thy-neighbor" policies—have emerged within the Region. With technical and professional advisory staffs paid for by federal and state grants, most of the county governments and many of the small units within the Region are embarked on formal programs to attract what, in their constituency's view, is desirable industrial and commercial development. Sometimes, these activities take the form of special commissions and research projects which review existing public policies for their effect on private locational decisions. Examples are the New Jersey Commission to Study Laws Affecting Industrial Development; the Long Island industrial survey carried out by a bureau of Hofstra College; and the studies sponsored by the New York State Department of Commerce.[13]

More frequently, the drive for industry becomes a major objective of county or local planning boards. Under their direction, land is set aside

---

[13] The New Jersey Commission's report was issued at Trenton in 1957. See also Hofstra College Bureau of Business and Community Research, *Long Island Industrial Survey* (Hempstead, 1956).

for future industrial development, transportation plans are made with a view to enhancing the area's industrial potential, and the private acquisition and improvement of land for business purposes are facilitated by public action. In all these endeavors, the governments are often closely associated with business and civic groups. Influential citizens are designated as "economic ambassadors." Special tours for outside industrialists are arranged to emphasize both the natural and political advantages of a given jurisdiction.

Not infrequently, these activities engender hostility among a municipality's neighbors, with accusations of "pirating" of industry or social irresponsibility. It was, for example, the promotional and zoning activities of Fairlawn, New Jersey, which led the Deputy City Administrator of New York in July 1959 to charge that the community which "had been luring industries out of New York is refusing to house workers from the city." In reply, New Jersey spokesmen stoutly denied any special concessions to industry. They did concede, however, that New Jersey's tax levels might be considered "assets" by a company on the move and that workers might find difficulty in buying homes in developments where "look alike" housing is not permitted and where the new homes being built were in the $20,000-and-up range.[14]

The application of zoning powers and promotional techniques varies substantially in different parts of the Region, and some of the instruments are far more effective than others. Quite frequently, comprehensive zoning regulations appear only after the pattern of development in a jurisdiction has been set, and the opportunities for controlling growth are thus reduced to fairly narrow limits. As for the mercantilist policies, there is little solid evidence that promotional efforts to attract new industry are effective counterweights to limitations imposed by geography or market considerations. Counterbalancing the Fairlawn story are studies in Long Island and Middlesex County, which report that special tax and service arrangements for new industry are often regarded with suspicion by a firm, and that it takes a fairly long time for a government to establish a reputation for a "favorable business climate."[15]

Nonetheless, planning, zoning, and promotion do represent ways by which all local units of general jurisdiction can keep "undesirables" out and encourage "desirables" to come in, if they choose. And of course the definition of desirables and undesirables varies from place to place. No common policy toward the control of land or the encouragement of economic development exists within the Region. Instead, land-use controls are applied in an atmosphere of intergovernmental jostling, and the ensuing

[14] "City Says Suburbs Bar Its Workers," *New York Times* (July 6, 1959).
[15] For example, Hofstra College, just cited, part I.

pattern of population and industrial distribution is often determined according to the respective political capacities of the municipalities involved.

The Appeal Upstairs

The policies and strategies we have summarized so far have all been largely the province of local governments. As such they are "Region-bound"—restricted in application to economic activities within the 22 counties and as varied in effectiveness as are the number and types of government. In these circumstances, a final estimate of their significance must await the later area-by-area survey, in which their effects can be localized.

The governments possess one more alternative, however, which is less dependent on their own particular mix of urbanization forces. By drawing on revenues from outside the Region, by turning to state and federal governments for support, the local units can to a considerable degree escape their environment. Indeed, they can find assistance of such dimensions as to dwarf the effects of their individual efforts to come to terms with urbanization. At least in terms of the additional sources of revenues, this avenue is of critical importance in accounting for the size and substance of the public sector.

For the Region as a whole, outside financial assistance is now second only to the property tax as a source of revenue. For several services, they are the prime support. In schools, welfare, health, highways, and redevelopment, outside contributions typically outweigh total local contributions for the respective services. Thus, in 1957, state aid was 20 per cent of the total revenue of New York City, 24 per cent of the total revenue of the local governments in the rest of the New York part of the Region, and 11 per cent for the New Jersey side. Broken down by type of government in 1955, the Region's general-purpose municipal governments except for New York City received over $21 million from state sources, compared to the $246 million raised by local property taxation. The county governments received more than $31 million compared to property revenue of $109 million. The school districts received $91 million as against $313 million in property levies.[16]

The experience of the New York and New Jersey sections of the Region and the experience of inner and outer counties within each state differ sharply in the patterns of grants-in-aid. Both in absolute and per capita terms, New York State has displayed a far greater disposition to put its funds at the disposal of its localities. Thus, in 1955, New York State

---

[16] U.S. *1957 Census of Governments,* Vol. VI, Nos. 28 and 30. As usual in this chapter, noncomparability of financial reporting units prevented our including Fairfield County in the Region totals.

aid to municipalities in the Region (again excepting New York City) was almost six times as great as that of New Jersey, while the total amount raised by property taxes in the New Jersey portion of the Region was twice that of the New York portion. In the period 1950–1955, the New York grants increased 34 per cent compared to a 9 per cent increase in New Jersey. On a per capita basis in 1955, state assistance to general governments below the county level amounted to 65 cents in the "New Jersey Inner" portion of the Region and $6.65 in "New York Inner." In the outer counties, the New Jersey units received grants averaging $1.24 per capita in contrast to the "New York Outer" average of $7.15.[17]

The numbers and types of assistance programs, as well as the size of the grants, serve as a further contrast between the two states. In New York, the array of state grants includes not only those primarily stimulated by the federal government in welfare, highways, and public health but also a series of programs supported almost entirely by the state. New York pioneered, for example, in the post-war development of general-purpose grants to municipalities—which for the whole state exceeded $108 million in 1956. Its educational assistance program is both sophisticated and comprehensive, covering some 22 special programs in addition to the basic "minimum foundation" grant which is aimed at establishing a floor for school expenditures across the state. Assistance is also offered in such varied fields as public works planning, service to veterans, conservation, probation, and delinquency. So large have the programs become that in 1956 state aid constituted over half of the state's annual operating expenditures.[18]

New Jersey takes a far more conservative posture. Statewide, four out of every five assistance dollars are earmarked for school purposes in New Jersey, and the bulk of the remaining allocations falls in programs strongly supported by the federal government. In 1958, state assistance to all school districts totaled $150.8 million. By contrast, the second largest grant program, welfare, amounted to $18.4 million, followed by highways at the $16.9 million mark; police and fire pensions, $4.4 million; and beach and waterways, $1.3 million. Allocations for all other programs totaled only $2.2 million—less than New York was assigning to Youth Service

[17] See note 9. See also New Jersey Taxpayers Association, *Notes on Government,* an occasional publication which details (as in Nos. 27, 28, 32) state grants-in-aid for major purposes.
[18] For a general description of the New York grant program see the mimeographed report of the State Department of Audit and Control, August 1956, entitled "State Aid to Local Government." The U.S. *1957 Census of Governments* provides comparative analysis, and an authoritative summary of the grant programs is found in Lynton K. Caldwell, *The Government and Administration of New York* (New York, 1954).

Bureaus. Perhaps even more significant is the comparatively sluggish trend in New Jersey grants over the past few years. Between 1953 and 1958, grants for welfare increased $4.4 million, highways $1 million, and fire and police pensions $3.4 million; beach and waterways declined $0.5 million; and all others increased $0.8 million. In vivid contrast, school assistance rose from $40.8 million to $107.6 million, an increase of 158 per cent. Indeed, between 1956 and 1958, school aid accounted for more than 90 per cent of the rise in all grant programs.[19]

The state-aid differences between New York and New Jersey emphasize—but by no means explain—the extraordinarily complex effect of the grant programs on the Region's political economy. One significant result of those differences is that New Jersey and New York municipalities equally endowed in economic resources and equally beset by public demands will be in quite unequal positions so far as their total resource base is concerned. Even more important, the formulas for allocating funds differ according to the legal status of the local governments involved, and therefore different volumes of support are found within the same state. In New York, the existence or nonexistence of a village within a town, the decision of a community to seek incorporation as a city, the kind of charter a county adopts—all affect the amount of state revenue available, quite apart from the number of residents or the assessed valuation within a given area.[20]

The total of federal and state money channeled into the New York Metropolitan Region is not, then, a reliable indicator of the contributions available in any given sector. Small jurisdictions share better proportionally than large; one kind of public function receives funds and another does not; two types of school district find their allocations calculated on the basis of quite different criteria. Though grants expand the tolerance limits in which local governments maneuver, these consequences are by no means uniform. Certainly, they do not have an equalizing influence on levels of expenditures.

The differential impact of the grant programs thus expands rather than reduces the range of political alternatives available to any single jurisdiction in coming to terms with its environment. Like the purely local strategies, grant programs provide further options for adjustment and additional revenues for the maintenance of the present system of metropolitan government. Since their effect cannot be estimated on an across-the-board

[19] New Jersey Taxpayers Association, *Notes on Government,* No. 32.
[20] For a general analysis of the effects of New York grants-in-aid upon governmental operations within the Region see Arch Dotson, "Metropolitan Aspects of New York State–New York City Fiscal Relations," a consultant memorandum in the 1956 report of the New York State–New York City Fiscal Relations Committee.

basis, the grants become subsumed as one of the array of weapons each jurisdiction has at its disposal.

Over time, then, each part of the Region has come to rely on a special combination of the strategies available and to fashion its particular style of adaptation. If we are to understand how the policies are actually applied—in contrast to what opportunities exist—we will have to abandon a general survey. In the next sections, we will build up piece-by-piece a knowledge of how each part of the Region has fashioned its policies of response: what room for maneuver exists, what the respective odds are for modifying or escaping the forces of urbanization, and how intense the infighting has become among local factions.

COMBINATIONS: THE NEW YORK SUBURBS— MULTIPLICATION AND DIVISION

In contrast to the City of New York, where officials appear prone to use whatever expedient lies at hand, the local governments in the seven New York counties outside the City still seem in a position to be selective in their strategies. To be sure, they are not free from financial problems; in some cases they are beset with difficulties perhaps exceeding that of the central city. But they have been in a better position to choose one or another major strategy with some promise of success.

For the longest-settled suburban communities, the strategy has been principally one of land-use control. Sometimes at the last moment before the bulldozers appear, sometimes with exceptional foresight, the residents of these suburbs have decided to resist the outward migration of people and business. In carrying out these policies, they have been assisted by the size and legal structure of their local governments. New York governments of general jurisdiction embrace far more geographical territory per unit than those in New Jersey. The county, in particular, has proved capable of centralizing many powers and programs of the kinds that, in New Jersey, are still in the hands of municipalities. The combination of large territory and power has permitted New York counties and towns a much broader and much more systematic use of planning and zoning powers than in New Jersey.

The outstanding example is, of course, Westchester County—and most particularly northern Westchester. At least since World War I, Westchester has been stereotyped as a refuge of upper-income families from the City who settle in "quality" neighborhoods and consequently enjoy high-quality public services with relatively low "tax effort." During the same period, the county's political leadership has devoted most of its energies to public policies which support the pattern of low densities which topography originally encouraged. Though exceptions exist among its municipalities, Westchester remains, as someone has quipped, dedicated "to zoning against 'Bronxification.' " The stand against "Bronxification" consists fundamentally of policies designed to maintain reasonably low levels of density; to exclude developments of a character likely to result in more public expenditures than they return in revenue; and to discourage the construction of major state highways wherever possible and, when not possible, to prevent unsightly commercial developments on through highways. More recently, the expansion of "Westchester-type" industry has been encouraged; outsiders have been excluded from the county's well-developed park system; and considerable attention has been paid to keeping out what has been termed "the undesirable element." In one expert's view "the exclusion policy will no doubt stick. There is only one state park in this county, Mohansic, and Westchesterians don't particularly want any more. . . . They do hope, of course, that Bob Moses will build lots of state parks elsewhere to draw off the pressure."[21]

The clear, if somewhat negatively expressed, image of what Westchester should be, supported by a tradition of political leadership which has enabled the county to stimulate timely municipal action, reveals itself in the sophisticated state of zoning and planning within the county. A check of eleven of the county's sixteen towns by the staff of the Regional Plan Association in 1958 revealed that in seven of them over 80 per cent of all residentially zoned property was zoned at one acre or more. Of the remaining four towns, one had 75 per cent zoned this way, and the others had approximately 50 per cent. Nor is it only the spaciously zoned communities which apply land-use controls. All 46 municipalities in Westchester had zoning ordinances in 1957—most of them operative for over a quarter of a century and four-fifths of them revised or newly adopted since 1945. The entire county's population is served by planning boards; over 90 per cent of the population live in areas where subdivision regulations apply and official maps are in existence; over 60 per cent are in jurisdictions where master plans have been adopted. A majority of the jurisdictions have retained planning consultants to carry on new planning, revise zoning ordinances, develop master plans, and advise on redevelopment projects.[22]

[21] Among the main sources for the description and analysis of Westchester are interviews with Merrill Folsom, the county's *New York Times* correspondent; William Cassella of the National Municipal League; and Robert T. Daland of the University of North Carolina. I have also had the opportunity to review Professor Daland's draft manuscript, "A Political System in Suburbia: The Politics of Autonomy," an authoritative treatment of the county. The quotations are taken from a letter from Daland to the author.
[22] Regional Plan Association files, and Westchester County, *The Status of Municipal Planning* (1957).

One result of the unflagging efforts to shape the county according to the vision of the first migrants from the City is the relative ease with which the county supports high-cost public services. Although Westchester in 1955 was third in total per capita expenditures among the seventeen counties of the Region outside New York City and had the highest expenditures among counties of about the same densities, its per capita increase in total property taxes between 1945 and 1955 was the lowest of all. In that period, its municipalities experienced an absolute decline in per capita payments for debt service—a record approached only by outlying Dutchess County.

In contrast, the other New York counties outside the City had increases ranging from 85 per cent to 228 per cent. Though revenue from non-property sources rose 68 per cent in Westchester, this trend too is below the Region's average when New York City is excluded. In short, while maintaining a high level of expenditures and providing services generally rated by professional observers as among the highest quality in the Region, the county units kept their public financial affairs in good order, and the proportion of their resources devoted to public purposes, whether measured in terms of income or market value of property, was below the average for the New York portion of the Region.[23]

Other local governments in the New York State portion of the Region, outside New York City, also rely on land-use policies to maintain a favorable political economy, though in no other county have such policies been so consistently and comprehensively applied. In Oyster Bay Township, Nassau County, eleven of the fifteen villages have zoned all land set aside for residential purposes at one acre or more. Here and there in North Hempstead there are municipal islands with similar regulations— Kings Point, North Hills, and Roslyn Harbor, for example. In these communities, the tradition of spacious life is so politically strong that their governments have turned aside the postwar wave of mass development which only a few miles south changed farmland into Levittowns. There are also hints that in counties yet to feel the full impact of the urban exodus—Rockland, Putnam, and possibly Suffolk and Orange—the Westchester lesson is being taken to heart. In the last ten years, planning boards have been activated, master plans adopted, and zoning laws enacted in an effort to produce "balanced" communities.[24]

Nevertheless, in none of the other New York counties has Westchester's zoning record been approached, and some have adopted altogether different strategies. In Nassau, for example, the fairly primitive land-use regula-

---

[23] See note 9.
[24] Regional Plan Association, *Bulletin 86* (1956). See also the *New York Herald Tribune's* two special feature series on Suburbia in April 1955 and April 1959.

tions existing at the end of World War II proved inadequate to stem the influx of new population in the center and south of the county. The rush eastward in search of vacant land set the stage for mass residential development, spurred on, in Robert Moses' acid words, by "the estate owner who is surrounded, can't pay higher taxes and has lost interest in his home and the community; the truck farmer who wants to retire or head east to cheaper open land; the speculator who aims to cut up real estate into as many postage stamp lots as weak zoning resolutions and weaker officials will permit." Reflecting on the recent past, Moses offered the judgment that "if intelligent forethought had forced larger lots and higher restrictions in recent subdivisions, the future overpopulation would have been controlled and most of the evils which flew out of that Pandora box would have been kept tightly under cover." Looking to the future, he could predict in blunt terms, "If lying or exaggerated real estate advertisements mean more to you than decent standards, if your surviving country squires continue to sell to developers for the most they can get and leave to jackals what they claimed to prize, if the small owner is so stupid that he permits cheap promoters to repeat the mistakes of the city, you are going to have suburban slums as sure as God made little apples."[25]

Mr. Moses' verdict on Long Island's possible future may be somewhat overdrawn—he himself has proclaimed that the joke is on the pessimist. Yet certainly in the decisive stage of its suburban settlement, Nassau's governments were not prepared to guide growth in the Westchesterian manner. On the contrary, its principal response has been to adapt the governmental structure to the new environment rather than to shape the character of the environment itself. One major modification has taken place in the county government: Nassau was the first New York county to adopt a "strong executive" plan. Its government has also assumed increasing responsibilities in the fields of highways, health, sewage construction, water supply, and welfare, and it has exercised a centralizing influence in tax assessment and collection. A countervailing structural development which has occurred, as population growth in specific parts of the county has generated specific pressures, has been the establishment of more and more special districts to handle specific public programs.

On balance, these latter forces have tended to predominate. While the number of villages in Nassau has remained at 63 since 1933 and the number of school districts has decreased from 65 to 62, a steady growth in other special districts has occurred. Between 1945 and 1955, their

[25] Robert Moses, "The Future of Nassau and Western Suffolk: Introductory Remarks," in *The Problems of Growth in Nassau and Western Suffolk: A Planning Forum, Hofstra College* (Hempstead, 1955), p 3.

number rose from 173 to 268, providing such diverse functions as fire protection, street lighting, parking, parks, police, garbage disposal, sanitary regulation, sewage facilities, water, and drainage.[26] It would be a mistake to regard the districts as totally unrelated to other governmental units, for they are created under town law, their tax and assessment policies are reviewed by the County Board of Supervisors, and their activities are knit together by informal political ties among their officials. What the districts do represent, however, is a device for gaining access to tax revenue which under state limitations might otherwise be foreclosed and for assigning particular tax bills to particular classes of inhabitants.

Thus the resident of Nassau County receives his services from, and pays his taxes to, a number of governments. They provide him with a plethora of opportunities to enjoy the benefits of home rule, but they also bring a considerable layering of tax bills and a considerable differential in taxes depending on residential location. As Samuel F. Thomas has pointed out, a property owner in one unincorporated area in the Town of Hempstead received in 1955 a consolidated town tax bill of $3.63 per $100 of assessed valuation representing the charges of ten separate districts, to which was added a second bill of $4.26 per $100 for schools, or a total of $7.89. By contrast, a property owner in an incorporated area received a consolidated town bill of $1.50 per $100 representing the charges of the county and town for general purposes and of the county sewer district, a school tax bill of $2.78, and a village tax bill of $1.79. Thus, his total tax bill was $6.07 per $100, almost $2 per $100 less than the resident in the unincorporated area. Although the standards of services were not necessarily the same, it is doubtful that services in the unincorporated area were greater in number or higher in professional quality. More likely, the reverse condition was true.[27]

It is not only at the town and village level that the principle of multiplication is applied in Nassau County. For the county and indeed for all of Long Island, special authorities and state commissions play an important role. Thus the Nassau County Bridge Authority was created in 1945 to acquire the privately owned Atlantic Beach Bridge and to construct a new span. Both the Jones Beach State Parkway Authority and the Bethpage Park Authority have important responsibilities in the construction of the county's highway network and the provision of recreation facilities. Tied to the Long Island State Park Commission by identical membership

[26] Samuel F. Thomas, "Nassau County: Its Governments and Their Expenditure and Revenue Patterns," draft report of a project of the New York Area Research Council of the City College of New York (mimeographed, 1957), p. 24.
[27] *Ibid.*, Part II, Section B.

on their governing boards, these three authorities provide another means of relieving the general government of responsibilities and costs.

The results of the strategy of fragmentation in Nassau is a public sector exceeded by no other in the Region in the resources diverted from the private sector. Nassau not only experienced by far the fastest growth in expeditures after the war—some 160 per cent between 1945 and 1955—but its per capita total expenditure in 1955 was $244 compared to the City's $226 and Westchester's $227. As the Commission on New York City–State Relations has pointed out, Nassau's total expenditures in 1954 amounted to 6.7 per cent of personal income, compared to 6.5 per cent and 5.6 per cent for New York City and Westchester respectively.[28]

Comparisons between expenditures and income give a sense of the magnitude of Nassau's public programs; they do not, however, indicate who pays the bills. It is especially important to realize that one out of every four residents is a child. Therefore the major local expenditure for Nassau is for public schools. And schools are more heavily supported by state aid than any other activity. In terms of property taxes per capita, for example, the amount collected in Nassau in 1955 was $128 in contrast to Westchester's $133.[29] In short, Nassau's high spending record is not necessarily accompanied by an equivalent drain on local resources. Its governments have managed to live with the consequences of expansion— although few of them may consider it gracious living.

The remainder of the New York counties do not present the sharp delineation in choice of weapons which distinguishes Westchester and Nassau. Suffolk, for example, has experienced the effects of the suburban migration only in its western areas and to date has behaved in its political response like a house divided. Its five western townships, within commuter range of the City, have increasingly sought to adopt some of Westchester's and Nassau's techniques. Its five eastern townships, as yet largely unaffected by the course of the Region's development, have continued on accustomed ways. The result, as described by Long Island's *Newsday,* is a "case of governmental rickets . . . brought about by a split personality between the rising suburban areas which need a strong county government . . . and political thinking in the east end which generally rejects the need for a county government any more modern than 1900."[30]

Whatever the root causes of the east-west split, Suffolk's bifurcated

[28] See note 20.
[29] See note 9.
[30] *Newsday,* series entitled "Suffolk Needs a Charter," article of Aug. 13, 1957. See also Suffolk County Planning Board, "Progress Report" (Patchogue, 1957).

politics has meant—at least until recently—that the county's response to the problems of growth has been more reflex than calculated. As development has proceeded, Suffolk has relied on either surplus funds husbanded from the past or large increases in assessed valuations of residential property. The efficacy of this policy seemed dubious in 1957 when "county officials became aware of a startling fact. Building was beginning to fall off. The county's assessed valuation appeared to be nearing a static point—leveled off by industry's unwillingness to establish itself in a county which has little to offer in the way of tax abatement programs, good roads, or a good labor pool. In the meantime, the county had obligated itself for more than $12,000,000 worth of new projects." To meet this situation, "the county board authorized a $50,000 special census. By proving a substantial rise in population, the county was able to claim an additional $1,000,000 in state aid."[31]

The pattern of instinctive reaction has varied widely among the Suffolk townships. Each of the ten towns until 1958 made its own assessments, with resulting variations from 23 to 43 per cent in the ratio of assessed to market value. Each had its own planning and zoning philosophy. According to Newsday, "four east end towns do not even have zoning ordinances and there is nothing to prevent the building of a glue factory next to someone's home. . . . Only one township has what is considered a proper allowance for industrial growth. . . . In Huntington, recently, complicated zoning requirements set off a legal tussle when the zoning board was haled into court by the town board. . . . In case after case, Suffolk towns, operating under 20-year zoning ordinances, have failed to stay ahead of or even keep up with fast-changing conditions. Suffolk zoning remains a hodge-podge of inconsistencies."[32]

There are signs that Suffolk is girding itself for the adoption of a more consistent policy toward growth. After two false starts, the county has adopted a new charter modeled after the Westchester and Nassau ones and providing for an elected county executive to begin unifying the county, administratively and psychologically. Even earlier the Suffolk County Planning Board had been revitalized and town planning and zoning activities speeded up. Inherent in this process of reorganization is a philosophy becoming more and more explicit, that Suffolk must rely on industrial development to provide the basis for its political economy. Thus Babylon has set aside 8 per cent of its land area for industrial use; Islip has undertaken to rezone 4000 acres; Huntington is planning to make industrial sites available along the new Long Island Expressway to Riverhead. Similarly, Greenport has offered an existing two-story building free to an ac-

[31] Newsday, Aug. 8, 1957.
[32] Ibid., Sept. 9.

ceptable business; Brookhaven has established a Town Industrial Advisory Committee; and Port Jefferson has been opened as a deep-sea port after a "three year battle against civic and boating groups."[33]

Economically, of course, one can question the feasibility of Suffolk's plans for industrial expansion. In comparison with other parts of the Region, this distant county continues at a disadvantage in transportation costs. Nevertheless, to the degree that a conscious strategy is emerging, this strategy aims at attracting business.

While this general policy is being fashioned, the comparatively moderate rate of Suffolk's growth relative to Nassau's, and the large amount of land still available, keep its public financial problems in manageable proportions. Suffolk's population doubled from 276,000 to 528,000 between 1950 and 1957. But, in per capita expenditures in 1955, it was fourth among the seven New York counties outside New York City. In the rate of growth in per capita expenditures between 1945 and 1955, it was fourth, and in the ratio of real estate taxes to income it was still fourth.[34]

Political strategies in the Region's other New York counties, Rockland, Putnam, Orange, and Dutchess, present a more muddled picture. In all of these counties, assessment ratios still provide ample room for maneuverability. In all of them, particularly Rockland, planning and zoning programs are being actively developed. Nonetheless all but six of the New York municipalities without zoning in 1957 were in these counties, and agitation for cooperative planning and uniformity of policy do not at this time seem as intense as in Suffolk. The northern counties have much lower densities than those prevailing in the other parts of the Region and much more vacant land. Their search for additional sources of revenue is not as evident, and their impetus toward expansion of services and facilities is more muted.

But the more isolated position of these four counties does not necessarily mean lower public expenditures on a per capita basis. On the contrary, Putnam in 1955 ranked first in this respect among the seven New York counties outside New York City, in contrast to Rockland, Orange, and Dutchess, which fell at the bottom of the New York list. In rate of change from 1945 to 1955, Putnam was second only to Nassau, and Rockland followed in third position.[35]

Putnam's high expenditure level appears to be in part a function of sparsity of settlement, for extremely low densities like extremely high densities make for high per capita expenditures. Undoubtedly, a second influence is the number of high-income families that have migrated there,

[33] *Ibid.,* Sept. 9
[34,35] See note 9.

purportedly to re-establish the Westchester of the twenties. Thus, the county seems presently to have the public services typical of rural areas, supported by the higher incomes of its newer residents. Rockland seems much more a part of the metropolitan area. It has a low density, to be sure, but its growth has passed beyond the break-point between rural and urban living, and the "get going" process of modern public programs has now appeared.

The same comparative ranking of these four northern counties exists so far as local taxes are concerned. Putnam was first again among the seven New York counties outside New York City in per capita local property taxes in 1955 while the other three bring up the rear—separated from the nearest county by a substantial $20 per capita.[36] This ranking bespeaks the mixed situation of peripheral counties where the array of public services need not be as elaborate as that in more urbanized areas, but where either a decision for high-quality services or the "diseconomies" of a rural environment may call forth heavy per capita public spending.

At this stage of incipient development, then, it is not surprising that the policies of the counties have not yet crystallized. Rockland, faced with the most immediate prospect of heavy immigration, is characterized, in the rueful language of a nameless developer, by "the politics of the mau-maus"—residents attached to rural bliss who descend from the hills to fight at the first hint of large-scale development.[37] Orange has not yet experienced the political battles of the Rockland variety. Dutchess is oriented as much to Poughkeepsie as to the center of the New York Metropolitan Region. Putnam has a mix of farmers and exurbanites living in an uneasy political peace.

If there is a single instrument on which these governments rely, it is the beneficence offered by the state through grant programs which are designed to favor their semirural status. In nonschool grants to municipalities in 1955, calculated on a per capita basis, the municipalities of Orange, Dutchess, and Putnam ranked first, second, and third among all the counties of the Region, and Rockland ranked fifth. In the five years between 1950 and 1955, the increase in such grants exceeded 20 per cent in Putnam and Rockland, and was 16 per cent in Orange. Dutchess experienced a moderate rise of 4 per cent. In the same period, Nassau experienced a *decrease* of 17 per cent, while she underwent her most rapid period of growth.[38]

When one reviews the experience of each New York county, then, a sequence of governmental policy-making appears, corresponding to the

[36] See note 9.
[37] *New York Herald Tribune* (April 26, 1959).
[38] See note 9.

stages of urbanization in which the jurisdictions find themselves. In the first onrush of urban settlement, a government may choose to face the mounting tide of population primarily by catering to the public needs of the new residents. If so, it may be compelled to expand its tax base, either by closing the gap between assessed and true property valuation or by seeking new revenue sources. Or it may create new institutions which have the legal right to call upon the wealth and income of the constituency a second or third time, or even more. Alternatively, a jurisdiction may anticipate the forces poised to enter its boundary lines and by timely planning or zoning divert the pressures of growth. It may even try to assure by promotional activities, land-use measures, and tax policy that population increases are accompanied by additions to the industrial base of the community.

Depending on which decision is made in the beginning, a basis is laid for new strategies at the more mature stage of development. Governments which have relied upon zoning and planning tend to adopt policies akin to "holding actions" and take on more and more the appearance of islands amid the surge of growth around them. Governments which have permitted their land to become densely populated intensify their search for new funds and may become immersed in technical maneuvers to expand debt limits and tax yields. As obsolescence sets in, the need for renewal becomes increasingly apparent and as the cost of renewal becomes recognized, appeals for new revenue sources or more grants-in-aid are mounted with growing intensity—though customarily they receive a cool reception at the capitol.

This pattern may not continue. Though Nassau and Westchester provide contrasting object lessons and the counties farther out have displayed in the 1950's an increasing awareness of the implications of growth, their decisions are not yet firm. Zoning and planning techniques which served Westchester well are not as applicable today. The real probability in the suburbs of the future is not one-acre lots of the relatively well-to-do, or the Levittowns of Nassau, but instead fairly compact developments interspersed with open spaces and industrial parks in the new style of Sterling Forest, in Orange County. Since a planning philosophy directed to that end is still in the process of formulation, one can doubt its application in advance of the migration itself.

CONSEQUENCES

What is the upshot of this complex process of move and countermove among the local governments of the Region, the net effect of the combinations of policies which the jurisdictions have adopted toward the increasing pressures of urbanization? Beyond providing an understanding of how differences in expenditure and revenue patterns arise apart from environmental circumstances, do the political maneuvers alter in any major way the underlying forces which shape the substance of the public sector? Do they have any really significant implications for the private sector?

Answers to these questions can necessarily be only conditional ones. In an estimate of the influence of the strategies, alone or in combination, in any particular part of the Region, we are in effect asking ourselves what would have occurred if the economic and population forces had operated without political intervention. What density pattern might have resulted or what would have been the revenue potential of a "natural" pattern of industrial location on the basis of its actual market value? We can simulate some of these conditions, but we can never examine them first-hand.

On the basis of the evidence which is on hand, however, certain general conclusions seem warranted. So far as the policies of redefining urban environments are concerned, these strategies, in and of themselves, have little effect on the mainstream of economic development. Their alert and timely application may save a politician's career, but as the process of urbanization wears on, the capacity for maneuver, at least on the basis of the property tax system, is sharply reduced. Municipalities also reach a limitation in the use of special districts whose effect is to increase the volume of the property revenues taken from a single geographical area. Here too the limitation of the size of the property tax base comes to apply. Thus, the focus of political attention turns, as in New York City, increasingly toward a search for new sources of revenue—a search the success of which depends on the vigor of the private sector itself.

So far as policies designed to guide land use are concerned, these can obviously be important. Westchester stands as an example of how effective political action, if applied with energy and imagination, can affect the pattern of population settlement and industrial location. And the enclaves in Nassau and parts of New Jersey similarly testify to the capacity of determined constituencies to thwart the natural pressures of expansion.

The real effectiveness of land-use policies hinges on their timing: the date when comprehensive programs are applied. In Nassau, zoning and planning measures were developed largely after the heaviest waves of migration had settled in. In New Jersey, the small size of the municipalities has precluded a comprehensive approach. In most of the Region, the economic gains promised to individual landowners by selling out their holdings have outweighed considerations of organized community development. For the future, one must conclude that in Suffolk, Rockland, Monmouth, and Middlesex, the counties most likely to receive the bulk of new growth, the issue of whether public policy will be decisive is still unclear.

As for the industrial development programs of various localities, they appear to have a Don Quixote quality of impracticability. Undoubtedly, particular firms have been influenced in their choice of sites by the discovery of a "friendly" government. But . . . tax levels do not appear to be prime considerations in the location of industry. Certainly the most ardent booster plan cannot compensate for a municipality's deficiencies in transportation, labor, character of terrain, and water.

Even grants-in-aid from higher levels of government, the major strategy of adjustment on which the Region's governments rely for survival, do not seem substantially to affect the process of economic development. Present grant formulas are not geared to compensate decisively for differences in either demand-oriented or supply-oriented dimensions of municipal environment and they leave the relative position of the governments undisturbed. Hence their impact is to exaggerate present differences in financial status—enabling the smaller, outlying jurisdictions to rock along under their present structure of organization and finance.

Not one of these strategies, then, has important implications for the private sector of the Region taken as an entity. An industry barred from one locality can in all probability find a hospitable reception in another with equivalent economic advantages. High-income families take refuge in Westchester, southern Putnam, and Fairfield, while mass developers make breakthroughs in Nassau or Monmouth or Rockland to provide middle-class housing. With so many different constituencies, many options are open for firms and households alike, and though the process of industrial and population diffusion may occasionally be skewed, the forces are not, in general, thwarted, turned aside, or guided.

Yet, if these policies have little effect in shaping the Regional economy, they do keep the local-government system continuously agitated *about* economic affairs. Indeed, they engender a pattern of behavior more closely approximating rivalries in world economic affairs than a domestic system of government intent on aiding the processes of economic development. Because particular combinations of strategies may be effective for any one jurisdiction, there is a strong tendency for each to "go it alone" to develop appropriate protective devices to escape the expensive public by-products of the private process of development. Municipalities come to concentrate on ways and means of getting as large a slice of the existing economic pie as possible and of mitigating the effects of new residential settlement. The development of hundreds of separate policies, in various combinations, among hundreds of jurisdictions engenders a spirit of contentiousness and competition. As the possibilities of shifting burdens within a municipality diminish, as development programs fail to counteract the economic considerations which predominate in locational decisions, as urbanization goes on apace, the temptation to embark on municipal mercantilism becomes stronger. Paradoxically, the policies also become less effective, since a government's neighbors are likely to adopt comparable tax rates, make the same zoning policy, and grant equal concessions to new industries. In these circumstances, the management of the political economy goes forward in ways localized, limited, and largely negative in character.

ROBERT L. LINEBERRY

EDMUND P. FOWLER

Reformism and Public Policies in American Cities

A decade ago, political scientists were deploring the "lost world of municipal government" and calling for systematic studies of municipal life which emphasized the political, rather than the administrative, side of urban political life.[1] In recent years, this demand has been generously answered and urban politics is becoming one of the most richly plowed fields of political research. In terms originally introduced by David Easton,[2] political scientists have long been concerned with inputs, but more recently they have focused their attention on other system variables, particularly the political culture[3] and policy outputs of municipal governments.[4]

The authors are indebted to Professors Robert T. Daland, James W. Prothro, William R. Keech, and James Q. Wilson for comments on an earlier draft of this paper. For assistance in statistical and methodological questions, the advice of Professor Hubert Blalock and Mr. Peter B. Harkins has been invaluable. The authors, of course, assume responsibility for all interpretation and misinterpretation.

[1] Lawrence J. R. Herson, "The Lost World of Municipal Government," *American Political Science Review*, LI (June, 1957), pp. 330–345; Robert T. Daland, "Political Science and the Study of Urbanism," *American Political Science Review*, LI (June, 1957), pp. 491–509.

[2] David Easton, "An Approach to the Analysis of Political Systems," *World Politics*, IX (April, 1957), pp. 383–400.

[3] Edward C. Banfield and James Q. Wilson, *City Politics* (Cambridge: Harvard University Press, 1963); see also James Q. Wilson and Edward C. Banfield, "Public-Regardingness as a Value Premise in Voting Behavior," *American Political Science Review*, LVIII (December, 1964), pp. 876–887.

[4] See, for example, Thomas R. Dye, "City-Suburban Social Distance and Public

The present paper will treat two policy outputs, taxation and expenditure levels of cities, as dependent variables. We will relate these policy choices to socioeconomic characteristics of cities and to structural characteristics of their governments. Our central research concern is to examine the impact of political structures, reformed and unreformed, on policymaking in American cities.

POLITICAL CULTURE, REFORMISM, AND POLITICAL INSTITUTIONS

The leaders of the Progressive movement in the United States left an enduring mark on the American political system, particularly at the state and municipal level. In the states, the primary election, the referendum, initiative and recall survive today. The residues of this *Age of Reform,*[5] as Richard Hofstadter called it, persist in municipal politics principally in the form of manager government and at-large and nonpartisan elections. The reformers were, to borrow Banfield and Wilson's phrase, the original embodiment of the "middle class ethos" in American politics. They were, by and large, White Anglo-Saxon Protestants reacting to the politics of the party machine, which operated by exchanging favors for votes.[6]

It is important that we understand the ideology of these reformers if we hope to be able to analyze the institutions which they created and their impact on political decisions. The reformers' goal was to "rationalize" and "democratize" city government by the substitution of "community oriented" leadership. To the reformers, the most pernicious characteristic of the machine was that it capitalized on socioeconomic cleavages in the population, playing on class antagonisms and on racial and religious differences. Ernest S. Bradford, an early advocate of commission government with at-large elections, defended his plans for at-large representation on grounds that

. . . under the ward system of governmental representation, the ward receives the attention, not in proportion to its needs but to the ability of its representatives to 'trade' and arrange 'deals' with fellow members. . . . Nearly

---

Policy," *Social Forces,* IV (1965), pp. 100–106; Raymond Wolfinger and John Osgood Field, "Political Ethos and the Structure of City Government," *American Political Science Review,* LX (June, 1966), pp. 306–326; Edgar L. Sherbenou, "Class, Participation, and the Council-Manager Plan," *Public Administration Review,* XXI (Summer, 1961), pp. 131–135; Lewis A. Froman, Jr., "An Analysis of Public Policies in Cities," *Journal of Politics,* XXIX (February, 1967), pp. 94–108.
[5] Richard Hofstadter, *Age of Reform* (New York: Alfred A. Knopf, 1955).
[6] John Porter East, *Council Manager Government: The Political Thought of Its Founder, Richard S. Childs* (Chapel Hill: University of North Carolina Press, 1965), p. 18.

every city under the aldermanic system offers flagrant examples of this vicious method of 'part representation.' The commission form changes this to representation of the city as a whole.[7]

The principal tools which the reformers picked to maximize this "representation of the city as a whole" were the commission, and later the manager, form of government, the nonpartisan election and the election at-large. City manager government, it was argued, produced a no-nonsense, efficient and business-like regime, where decisions could be implemented by professional administrators rather than by victors in the battle over spoils. Nonpartisan elections meant to the reformer that state and national parties, whose issues were irrelevant to local politics anyway, would keep their divisive influences out of municipal decision-making. Nonpartisan elections, especially when combined with elections at-large, would also serve to reduce the impact of socioeconomic cleavages and minority voting blocs in local politics. Once established, these institutions would serve as bastions against particularistic interests.

Banfield and Wilson have argued that the "middle class ethos" of the reformers has become a prevalent attitude in much of political life. The middle class stands for "public-regarding" virtues rather than for "private-regarding" values of the ethnic politics of machines and bosses. The middle class searches for the good of the "community as a whole" rather than for the benefit of particularistic interests.[8] Agger, Goldrich, and Swanson, in their study of two western and two southern communities, have documented the rise of a group they call the "community conservationists," who "see the values of community life maximized when political leadership is exercised by men representing the public at large, rather than 'special interests.' "[9] Robert Wood has taken up a similar theme in his penetrating analysis of American suburbia. The "no-party politics of suburbia" is characterized by "an outright reaction against partisan activity, a refusal to recognize that there may be persistent cleavages in the electorate and an ethical disapproval of permanent group collaboration as an appropriate means of settling disputes."[10] This ideological opposition to partisanship is a product of a tightly-knit and homogeneous community, for "nonpartisanship reflects a highly integrated community life with a powerful capacity to induce conformity."[11]

Considerable debate has ensued over both the existence and the conse-

[7] Ernest S. Bradford, *Commission Government in American Cities* (New York: Macmillan, 1911), p. 165.
[8] Banfield and Wilson, *op. cit.,* p. 41.
[9] Robert Agger, Daniel Goldrich, and Bert E. Swanson, *The Rulers and the Ruled* (New York: Wiley, 1964), p. 21.
[10] Robert C. Wood, *Suburbia: Its People and Their Politics* (Boston: Houghton Mifflin, 1959), p. 155.
[11] *Ibid.,* p. 154.

quences of these two political ethics in urban communities. Some evidence has supported the view that reformed governments[12] are indeed found in cities with higher incomes, higher levels of education, greater proportions of Protestants, and more white-collar job-holders. Schnore and Alford, for example, found that "the popular image of the manager city was verified; it does tend to be the natural habitat of the upper middle class." In addition, manager cities were "inhabited by a younger, more mobile population that is growing rapidly."[13]

More recently, Wolfinger and Field correlated socio-economic variables—particularly ethnicity and region—to political structures. They concluded that "the ethos theory is irrelevant to the South . . . inapplicable to the West . . . fares badly in the Northeast . . . " and that support for the theory in the Midwest was "small and uneven."[14] Region proved to be a more important predictor of both government forms and of policy outputs like urban renewal expenditures than did the socio-economic composition of the population.

In our view, it is premature to carve a headstone for the ethos theory. It is our thesis that governments which are products of the reform movement behave differently from those which have unreformed institutions, even if the socio-economic composition of their population may be similar. Our central purpose is to determine the impact of both socio-economic variables and political institutions (structural variables) on outputs of city governments. By doing this, we hope to shed some additional illumination on the ethos theory.

RESEARCH DESIGN

Variables

The independent variables used in this analysis, listed in Table 1, constitute relatively "hard" data, mostly drawn from the U.S. census.[15] These

[12] We refer to cities characterized by commission or manager government, nonpartisan elections, and at-large constituencies as "reformed." Our use of the term is historical and no value position on reformism's merits is intended. To refer to reformed cities as "public-regarding" or "middle class" is, it seems, to assume what needs to be proved.

[13] Leo Schnore and Robert Alford, "Forms of Government and Socio-Economic Characteristics of Suburbs," *Administrative Science Quarterly,* VIII (June, 1963), pp. 1–17. See also the literature cited in Froman, *op. cit.*

[14] Wolfinger and Field, *op. cit.,* pp. 325–326.

[15] The source for the first nine variables is *The City and County Data Book* (Washington: United States Bureau of the Census, 1962). For the last three variables, the source is Orin F. Nolting and David S. Arnold (eds.), *The Municipal Yearbook, 1965* (Chicago: International City Managers' Association, 1965), p. 98 ff.

variables were selected because they represent a variety of possible social cleavages which divide urban populations—rich vs. poor, Negro vs. White, ethnic vs. native, newcomers vs. old-timers, etc. We assume that such social and economic characteristics are important determinants of individual and group variations in political preferences. Data on each of these independent variables were gathered for each of the two hundred cities in the sample.[16]

TABLE 1 INDEPENDENT VARIABLES

1. Population, 1960
2. Per cent population increase or decrease, 1950–60
3. Per cent non-white
4. Per cent of native population with foreign born or mixed parentage
5. Median income
6. Per cent of population with incomes below $3000
7. Per cent of population with incomes above $10,000
8 Median school years completed by adult population
9. Per cent high school graduates among adult population
10. Per cent of population in white collar occupations
11. Per cent of elementary school children in private schools
12. Per cent of population in owner-occupied dwelling units

Our principal theoretical concern will be with the consequences of variations in the structural characteristics of form of government, type of constituency, and partisanship of elections. The variable of form of government is unambiguous. Except for a few small New England towns, all American cities have council-manager, mayor-council, or commission government. There is, however, somewhat more ambiguity in the classification of election type. By definition, a "nonpartisan election is one in which no candidate is identified on the ballot by party affiliation."[17] The legal definition of nonpartisanship conceals the wide variation between Chicago's and Boston's nominal nonpartisanship and the more genuine variety in Minneapolis, Winnetka, and Los Angeles.[18] We will quickly see, though, that formal nonpartisanship is not merely an empty legal

---

[16] We used a random sample of 200 of the 309 American cities with populations of 50,000 or more in 1960. All information on the forms of government and forms of election are drawn from *The Municipal Yearbook, 1965, op. cit.*
[17] Banfield and Wilson, *op. cit.*, p. 151.
[18] For Minneapolis, see Robert Morlan, "City Politics: Free Style," *National Municipal Review*, XLVIII (November, 1949), pp. 485–490; Winnetka, Banfield and Wilson, *op. cit.*, p. 140; Los Angeles, Charles G. Mayo, "The 1961 Mayoralty Election in Los Angeles: The Political Party in a Nonpartisan Election," *Western Political Quarterly*, XVII (1964), pp. 325–339.

nicety, but that there are very real differences in the political behavior of partisan and nonpartisan cities, even though we are defining them in legal terms only.[19]

Our classification of constituency types into only two groups also conceals some variation in the general pattern. While most cities use either the at-large or the ward pattern of constituencies exclusively, a handful use a combination of the two electoral methods. For our purposes, we classified these with district cities.

The dependent variables in this study are two measures of public policy outputs. A growing body of research on local politics has utilized policy measures as dependent variables.[20] The present research is intended to further this study of political outputs by relating socio-economic variables to expenditure and taxation patterns in cities with varying political structures.

The dependent variables are computed by a simple formula. The measure for taxation was computed by dividing the total personal income of the city into the total tax of the city, giving us a tax/income ratio. Similarly, dividing expenditures by the city's aggregate personal income gave us an expenditure/income ratio as the measure for our second dependent variable. These measures, while admittedly imperfect,[21] permit us to ask how much of a city's income it is willing to commit for public taxation and expenditures.

Hypothesis

Much of the research on city politics has treated reformed institutions as dependent variables. Although we shall briefly examine the social and

[19] At least one other variable may produce a given institutional form in a city—the legal requirements of a state government, which vary from state to state and may even vary for different kinds of cities within the same state. We have not taken account of this variable because systematic information on comparative state requirements in this area was unavailable to us. However, Wolfinger and Field consulted several experts and eliminated cities which are not given free choice over their institutions. Nevertheless, a comparison of our figures with theirs revealed no important differences.

[20] See footnote 4, *supra*.

[21] We recognize that these are only rough indicators of city finance policies. Definitions of taxation vary from city to city and what may be financed from taxes in one city may be financed from fees in another. Expenditures present a more complex problem because the types and amounts of state transfer payments vary from state to state according to state laws, the division of government labor in a state, the incomes and sizes of cities, not to mention political factors at the state level. We think it important, however, that our independent variables explain a large proportion of the variation in municipal outputs as we measured them. No doubt one could explain an even larger proportion of the variation in measures which specify different functional responsibilities of cities. At least these measures constitute a starting point, and we hope others will improve on them. The source of our output measures was the *City and County Data Book, op. cit.*

economic differences between reformed and unreformed cities, our principal concern will be to explore the *consequences* for public policy of political institutions. From our earlier discussion of the political culture of cities, we hypothesized that:

1. The relationship between socio-economic cleavages and policy outputs is stronger in unreformed than in reformed cities.

This hypothesis focuses on the intention of the reformers to minimize the role of particularistic interests in policy making.

REFORMED AND UNREFORMED CITIES: A COMPARISON

The economic and social contrasts between reformed and unreformed cities have been the subject of much research,[22] and for our purposes we may be brief in our treatment. We divided independent variables into three groups, one measuring population size and growth, a second containing social class indicators, and a third including three measures of social homogeneity. The means and standard deviations for each variable by institutional category are found in Table 2.

It should initially be noted that population size and growth rate fairly clearly separate the reformed from the unreformed cities. As Alford and Scoble have amply documented,[23] the larger the city, the greater the likelihood of its being unreformed; the faster its growth rate, the more likely a city is to possess manager government, nonpartisan and at-large elections. These differences are largely accounted for by the fact that very large cities are most likely to (1) have unreformed institutions and (2) be stable or declining in population. Since neither of these variables emerged as particularly important predictors of our output variables, we relegated them to secondary importance in the rest of the analysis.

The data in Table 2 indicate that reformed cities (at least those over 50,000) do not appear to be "the natural habitat of the upper middle class." While reformed cities have slightly more educated populations and slightly higher proportions of white collar workers and home ownership, unreformed cities have generally high incomes. In any case, whatever their direction, the differences are not large. What is striking is not the differences between the cities but the similarities of their class composition.

[22] See, for example, Robert Alford and Harry Scoble, "Political and Socio-Economic Characteristics of American Cities," *The Municipal Yearbook, 1965, op. cit.,* pp. 82–97; Sherbenou, *op. cit.;* John H. Kessel, "Governmental Structure and Political Environment," *American Political Science Review,* LVI (September, 1962), pp. 615–620.

[23] Alford and Scoble, *op. cit.* The particularly large differences found between the populations of reformed and unreformed cities reflect the fact that New York City and several other urban giants are included in the sample.

TABLE 2 COMPARISON OF THE MEANS (AND STANDARD DEVIATIONS) OF
SOCIO-ECONOMIC CHARACTERISTICS OF REFORMED AND
UNREFORMED CITIES

Independent Variable	Government Type		
	Mayor-Council	Manager	Commission
Population			
Population (10^3)	282.5 (858.6)	115.7 (108.0)	128.6 (115.2)
Per cent change, 1950–1960	36.4 (118.8)	64.1 (130.4)	18.5 (36.7)
Class			
Median income	$6199 (1005.0)	$6131 (999.6)	$5425 (804.4)
Per cent under $3000	15.3 (7.0)	17.3 (6.9)	21.5 (7.9)
Per cent over $10,000	16.9 (7.2)	17.5 (6.7)	12.5 (3.7)
Per cent high school graduates	40.7 (10.8)	48.1 (8.9)	41.6 (10.4)
Median education (yrs.)	10.7 (1.1)	11.4 (.89)	11.0 (2.1)
Per cent owner-occupied dwelling units	54.9 (15.1)	57.3 (13.6)	54.6 (13.7)
Per cent white collar	44.1 (9.0)	48.1 (7.1)	44.2 (7.6)
Homogeneity			
Per cent nonwhite	10.6 (11.5)	11.6 (10.8)	16.5 (14.9)
Per cent native with foreign born or mixed parentage	19.7 (9.9)	12.4 (8.3)	11.7 (10.7)
Per cent private school attendance	23.5 (11.9)	15.3 (11.8)	16.6 (11.8)
	$N = 85$	$N = 90$	$N = 25$

Independent Variable	Election Type	
	Partisan	Nonpartisan
Population		
Population (10^3)	270.8 (1022.1)	155.8 (198.7)
Per cent population increase 1950–1960	17.1 (40.1)	58.3 (136.1)
Class		
Median income	$5996 (904.5)	$6074 (1045.5)
Per cent under $3000	16.8 (7.1)	17.2 (7.2)
Per cent over $10,000	16.1 (6.1)	16.7 (7.0)
Per cent high school graduates	40.5 (9.2)	45.3 (10.6)

TABLE 2 (Continued)

	Election Type			
	Partisan		Nonpartisan	
Median education (yrs.)	10.6	(1.1)	11.2	(1.2)
Per cent owner-occupied dwelling units	51.5	(14.4)	57.7	(13.8)
Per cent white collar	43.5	(7.5)	46.7	(8.3)
Homogeneity				
Per cent nonwhite	13.0	(11.9)	11.5	(11.8)
Per cent native with foreign born or mixed parentage	17.5	(10.7)	14.7	(9.6)
Per cent private school attendance	24.1	(13.6)	16.9	(11.3)
	$N = 57$		$N = 143$	

Independent Variable	Constituency Type			
	District		At-Large	
Population				
Population (10^3)	246.9	(909.8)	153.6	(191.2)
Per cent population increase 1950–1960	23.1	(36.4)	59.1	(143.7)
Class				
Median income	$6297	(965.2)	$5942	(1031.9)
Per cent under $3000	14.7	(6.5)	18.2	(7.6)
Per cent over $10,000	17.7	(7.1)	16.0	(6.6)
Per cent high school graduates	43.6	(10.9)	44.4	(10.4)
Median education (yrs.)	10.9	(1.1)	11.2	(1.2)
Per cent owner-occupied dwelling units	55.1	(14.4)	56.9	(14.5)
Per cent white collar	45.2	(9.4)	46.3	(7.5)
Homogeneity				
Per cent non white	9.8	(10.6)	13.0	(12.3)
Per cent native with foreign born or mixed parentage	18.9	(9.4)	13.4	(9.7)
Per cent private school attendance	23.2	(12.5)	16.6	(11.7)
	$N = 73$		$N = 127$	

Homogeneity is easily one of the most ambiguous terms in the ambiguous language of the social sciences. We have followed Alford and Scoble who used three measures of homogeneity: for ethnicity, the per cent of population native born of foreign born or mixed parentage; for race, the per cent nonwhite; and for religious homogeneity, the per cent of elementary school children in private schools. The last measure, while indirect, was the only one available, since data on religious affiliation are not collected by the Census Bureau.

With the exception of race, reformed cities appear somewhat more homogeneous than unreformed cities. While the differences in homogeneity are more clear-cut than class differences, this hardly indicates that reformed cities are the havens of a socially homogeneous population. Although the average nonpartisan city has 16.9 per cent of its children in private schools, this mean conceals a wide range—from 2 to 47 per cent.

Our findings about the insignificance of class differences between reformed and unreformed cities are at some variance with Alford and Scoble's conclusions. There is, however, some support for the argument that reformed cities are more homogeneous. While we used cities with a population of over 50,000, their sample included all cities over 25,000; and varying samples may produce varying conclusions. The only other study to analyze cities over 50,000 was Wolfinger and Field's and our conclusions are generally consistent with theirs. We differ with them, however, on two important questions.

First, Wolfinger and Field argued that what differences there are between unreformed and reformed cities disappear when controls for region are introduced: "The salient conclusion to be drawn from these data is that one can do a much better job of predicting a city's political form by knowing what part of the country it is in than by knowing anything about the composition of its population."[24] Since regions have had different historical experiences, controls for region are essentially controls for history and, more specifically, historical variation in settlement patterns. The problem with this reasoning, however, is that to "control" for "region" is to control not only for history but for demography as well: to know what region a city is in *is* to know something about the composition of its population. Geographical subdivisions are relevant subjects of political inquiry only because they are differentiated on the basis of attitudinal or socio-economic variables. The South is not a distinctive political region because two surveyors named Mason and Dixon drew a famous line, but because the "composition of its population" differs from the rest of the country.

It is therefore difficult to unravel the meaning of "controlling" for "region" since regions are differentiated on precisely the kinds of demographic variables which we (and Wolfinger and Field) related to reformism. Cities in the Midwest, for example, have a much higher proportion of home ownership (64%) than cities in the Northeast (44%), while Northeastern cities have more foreign stock in their population (27%) than the Midwest (16%). Hence, to relate ethnicity to political reformism and then to "control" for "region" is in part to relate ethnicity to reformism and

[24] *Op. cit.,* p. 320.

then to control for ethnicity. Consequently, we have grave reservations that the substitution of the gross and unrefined variable of "region" for more refined demographic data adds much to our knowledge of American cities. "Controlling" for "region" is much more than controlling for historical experiences, because region as a variable is an undifferentiated *potpourri* of socioeconomic, attitudinal, historical, and cultural variations.[25]

We also differ with Wolfinger and Field in their assertion that their analysis constitutes a test of the ethos theory. As we understand it, Banfield and Wilson's theory posits that particular attitudes are held by persons with varying sociological characteristics (ethnic groups and middle class persons, in particular) and that these attitudes include preferences for one or another kind of political institution. But relating the proportion of middle class persons in a city's population to its form of government says nothing one way or another about middle class preferences. An important part of understanding, of course, is describing and it is certainly useful to know how reformed cities differ from unreformed cities.

In our view, however, such tests as Wolfinger and Field used cannot logically be called explanations, in any causal sense. The most obvious reason is that they violate some important assumptions about time-order: independent variables are measured with contemporary census data, while the dependent variables are results of decisions made ten to fifty years ago. Moreover, this problem is multiplied by the difficulty of inferring configurations of political power from demographic data. Presumably, their assumption is that there is a simple linear relationship between sheer numbers (or proportions) of, say, middle class persons and their political power: the larger the size of a group in the city's population, the easier it can enforce its choice of political forms. At least one prominent urban sociologist, however, has found empirical support for precisely the opposite proposition. Hawley concluded that the smaller the proportion of middle class persons in a city, the greater their power over urban renewal policies.[26] Similarly, it may also be dubious to assume that the size of an ethnic population is an accurate indicator of influence of ethnic groups. Although we recognize the importance of describing the socioeconomic correlates of political forms, the logical problems in-

---

[25] In statistical parlance, the problem with "region" as an independent variable might be described as treating a complicated background variable as the first variable in a specific developmental sequence. But, as Blalock argues, ". . . *one should avoid complex indicators that are related in unknown ways to a given underlying variable.* Geographical region and certain background variables appear to have such undesirable properties." Hubert M. Blalock, *Causal Inferences in Nonexperimental Research* (Chapel Hill: University of North Carolina Press, 1964), p. 164 (italics in original).

[26] Amos Hawley, "Community Power and Urban Renewal Success," *American Journal of Sociology*, LXVIII (January, 1963), pp. 422–431.

volved suggest the need for a good deal of caution in interpreting these differences as explanations.[27]

In any case, the question of why the city adopts particular structures is of less interest to us than their consequences for public policy. It is to this analysis that we now turn.

POLICY OUTPUTS AND THE RESPONSIVENESS OF CITIES

We are now in a position to take three additional steps. First, we can compare the differences in policy outputs between reformed and unreformed cities. Second, we can assess the cumulative impact of socio-economic variables on these policy choices. Finally, we can specify what variables are related in what ways to these output variables. In essence, we can now treat political institutions, not as dependent variables, but as factors which influence the *level* of expenditures and taxation and the *relationship* between cleavage variables and these outputs.

Differences between Reformed and Unreformed Cities' Outputs

Contrary to Sherbenou's conclusions about Chicago suburbs,[28] our data indicate that reformed cities both spend and tax less than unreformed

TABLE 3   MEAN VALUES OF TAX/INCOME AND EXPENDITURE/
INCOME RATIOS, BY STRUCTURAL CHARACTERISTICS

	Per Cent	
Structural Variables	Taxes/Income	Expenditures/Income
Election type		
Partisan	.032	.050
Nonpartisan	.030	.053
Government type		
Mayor-council	.037	.058
Manager	.024	.045
Commission	.031	.057
Constituency type		
Ward	.036	.057
At-large	.027	.049

[27] See also the exchange between Banfield and Wilson and Wolfinger and Field in "Communications," *American Political Science Review,* LX (December, 1966), pp. 998–1000.

[28] Sherbenou, *op. cit.,* pp. 133–134.

cities, with the exception of expenditures in partisan and nonpartisan cities. It appears that partisan, mayor-council, and ward cities are less willing to commit their resources to public purposes than their reformed counterparts. What is of more importance than the difference in outputs, however, is the relative responsiveness of the two kinds of cities to social cleavages in their population.

The Responsiveness of Cities

We have argued that one principal goal of the reform movement in American politics was to reduce the impact of partisan, socio-economic cleavages on governmental decision making, to immunize city governments from "artificial" social cleavages—race, religion, ethnicity, and so on. As Banfield and Wilson put their argument, the reformers "assumed that there existed an interest ('the public interest') that pertained to the city 'as a whole' and that should always prevail over competing, partial (and usually private) interest."[29] The structural reforms of manager government, at-large, and nonpartisan elections would so insulate the business of governing from social cleavages that "private-regarding" interests would count for little in making up the mind of the body politic. But amid the calls of the reformers for structural reforms to muffle the impact of socio-economic cleavages, a few hardy souls predicted precisely the opposite consequence of reform: instead of eliminating cleavages from political decision-making, the reforms, particularly the elimination of parties, would enhance the conflict. Nathan Matthews, Jr. a turn-of-the-century mayor of Boston, issued just such a warning:

As a city is a political institution, the people in the end will divide into parties, and it would seem extremely doubtful whether the present system, however illogical its foundation be, does not in fact produce better results, at least in large cities, than if the voters divided into groups, separated by property, social or religious grounds.[30]

Matthews recognized implicitly what political scientists would now call the "interest aggregation" function of political parties.[31] Parties in a democracy manage conflict, structure it, and encapsulate social cleavages under the rubric of two or more broad social cleavages, the parties themselves. "Parties tend to crystallize opinion, they give skeletal articulation to a shapeless and jelly-like mass . . . they cause similar opinions to

[29] *Op. cit.,* p. 139.
[30] Quoted in Banfield and Wilson, *op. cit.,* p. 154.
[31] For a discussion of the concept of interest aggregation, see Gabriel Almond, "Introduction: A Functional Approach to Comparative Politics," in Gabriel Almond and James S. Coleman (eds.), *The Politics of Developing Areas* (Princeton: Princeton University Press, 1960), pp. 38–45.

coagulate . . ."[32] The parties "reduce effectively the number of political opinions to manageable numbers, bring order and focus to the political struggle, simplify issues and frame alternatives, and compromise conflicting interests."[33] Since parties are the agencies of interest aggregation, so the argument goes, their elimination makes for greater, not lesser, impact of social cleavages on political decisions.

Political scientists have recently confirmed Matthews' fears, at least with regard to electoral behavior in partisan and nonpartisan elections. Evidence points to the increased impact of socio-economic cleavages on voting when a nonpartisan ballot is used than when the election is formally partisan. Gerald Pomper studied nonpartisan municipal elections and compared them with partisan elections for the New Jersey State Assembly in Newark. He concluded that the "goal of nonpartisanship is fulfilled, as party identification does not determine the outcome. In place of party, ethnic affiliation is emphasized and the result is 'to enhance the effect of basic social cleavages.' "[34] If (1) this is typical of other American cities and if (2) electoral cleavages can be translated effectively into demands on the government in the absence of aggregative parties, then we might assume that the reformed institutions would reflect cleavages more, rather than less, closely than unreformed ones.

Essentially, then, there are two contrasting views about the consequences of municipal reform. One, the reformers' ideal, holds that institutional reforms will mitigate the impact of social cleavages on public policy. The other argues that the elimination of political parties and the introduction of other reforms will make social cleavages more, rather than less, important in political decision-making.

The Measurement of Responsiveness

We have hypothesized that socio-economic cleavages will have less impact on the policy choices of reformed than of unreformed governments. Thus, one could do a better job of predicting a city's taxation and expenditure policy using socio-economic variables in partisan, mayor, and ward cities than in nonpartisan, manager, and at-large cities. Operationally, we will test this hypothesis by using multiple correlation coefficients. Squaring these coefficients, called "multiple R's," will give us a summary measure of the total amount of variation in our dependent variables explained

---

[32] Maurice Duverger, *Political Parties* (New York: Science Editions, 1963). p. 378.
[33] Frank J. Sorauf, *Political Parties in the American System* (Boston: Little, Brown and Co., 1964), pp. 165–166.
[34] Gerald Pomper, "Ethnic and Group Voting in Nonpartisan Municipal Elections," *Public Opinion Quarterly,* XXX (Spring, 1966), p. 90; see also, J. Leiper Freeman, "Local Party Systems: Theoretical Considerations and a Case Analysis," *American Journal of Sociology,* LXIV (1958), pp. 282–289.

Independent Variables	Structural Variables (Per Cent)		Dependent Variable
	Reformed institution		
	Government: commission	62	
	Government: council manager	42	
	Election: nonpartisan	49	
	Constituency: at-large	49	
Twelve socio-economic variables			Tax/income ratio
	Unreformed Institution		
	Government: mayor-council	52	
	Election: partisan	71	
	Constituency: ward/mixed	59	

FIGURE 1   Proportion of variation explained (R^2) in taxation policy with twelve socio-economic variables, by institutional characteristics (in the total sample, the twelve independent variables explained 52% of the variation in taxes).

by our twelve independent variables.[35] The results of the correlation analysis are summarized in Figures 1 and 2.

On the whole, the results of the correlation analysis strikingly support the hypothesis, with the exception of commission cities. Thus, we can say, for example, that our twelve socio-economic variables explain 71 per cent of the variation in taxation policy in partisan cities, and 49 per cent of the variation in nonpartisan cities. In commission cities, however, socio-economic variables predict substantially more variation in both taxes and expenditures than in the unreformed mayor-council cities.[36] The anomaly of commission governments is an interesting one, for they present, as we will see, marked exceptions to virtually every pattern of relationships we found. The substantial explanatory power of these socio-economic

[35] It is possible that the difference between any two correlations may be a function of very different standard deviations of the independent variables. A quick look at Table 2, however, suggests that this is not likely to affect the relationships we find.

[36] Wolfinger and Field, *op. cit.,* p. 312, ". . . omit the commission cities from consideration since this form does not figure in the ethos theory." Historically, however, commission government was the earliest of the structures advocated by the Progressives and is quite clearly a product of the reform era. While history tells us that commission cities cannot legitimately be excluded from the fold of reformism, they appear to be its black sheep, characterized by low incomes, low population growth, and large proportions of nonwhites. In fact, they present a marked contrast to both mayor-council and manager cities.

Independent Variables	*Structural Variables (Per Cent)*	*Dependent Variable*
	Reformed institution	
	Government: commission	59	
	Government: council-manager	30	
	Constituency: at-large	36	
	Elections: nonpartisan	41	
Twelve socio-economic		Expenditure/income
variables		ratio
	Unreformed institution	
	Government: mayor-council	42	
	Constituency: ward/mixed	49	
	Elections: partisan	59	

FIGURE 2   Proportion of variation explained (R^2) in expenditure policy with twelve socio-economic variables, by institutional characteristics (in the total sample, the twelve independent variables explained 36% of the variation in expenditures).

variables is not altered, but confirmed, by examining the variables independently. The rest of the correlations show a consistent pattern: reformed cities are less responsive to cleavages in their population than unreformed cities.

If one of the premises of the "political ethos" argument is that reformed institutions give less weight to the "private-regarding" and "artificial" cleavages in the population, that premise receives striking support from our analysis. Our data suggest that when a city adopts reformed structures, it comes to be governed less on the basis of conflict and more on the basis of the rationalistic theory of administration. The making of public policy takes less count of the enduring differences between White and Negro, business and labor, Pole and WASP. The logic of the bureaucratic ethic demands an impersonal, apolitical settlement of issues, rather than the settlement of conflict in the arena of political battle.

TO SPEND OR NOT TO SPEND

If efforts to expand or contract the scope of government stand at the core of municipal political life,[37] they are nowhere better reflected than in the taxation and expenditure patterns of cities. A generation ago,

[37] Agger et al., *op. cit.*, pp. 4–14.

Charles Beard wrote that, "In the purposes for which appropriations are made the policies of the city government are given concrete form—the culture of the city is reflected. Indeed, the history of urban civilization could be written in terms of appropriations, for they show what the citizens think is worth doing and worth paying for."[38] Pressures to expand and contract government regulations and services are almost always reflected one way or another in the municipal budget. Labor, ethnic groups, the poor, and the liberal community may press for additional services and these must be paid for; the business community may demand municipal efforts to obtain new industry by paring city costs to create a "favorable business climate"; or businessmen may themselves demand municipal services for new or old business. In any case, few political conflicts arise which do not involve some conflict over the budget structure.

Class Variables and Public Policies

Part of the political rhetoric associated with the demand for a decrease in the scope of the national government is the argument that the initiative for policy-making should rest more with the state and local governments. Opposition to federal spending levels, as V. O. Key has demonstrated, is found more often among persons with middle class occupations than among blue-collar workers.[39] It is not inconceivable that the middle class argument about state and local responsibility might be more than political rhetoric, and that at the local level, middle class voters are willing to undertake major programs of municipal services, requiring large outlays of public capital. Wilson and Banfield have argued that the "public-regarding" upper-middle class voters in metropolitan areas are often found voting for public policies at variance with their "self-interest narrowly conceived," and that "the higher the income of a ward or town, the more taste it has for public expenditures of various kinds."[40] Similarly a longitudinal study of voting patterns in metropolitan Cleveland found that an index of social rank was positively correlated with favorable votes on welfare referenda.[41] If these data reflect middle-class willingness to spend on a local level, they might indicate that the "states' rights" argument was more than ideological camouflage: middle class voters stand foursquare behind public expenditures at the local level even when they oppose those

[38] Charles A. Beard, *American Government and Politics* (New York: Macmillan, 1924, 4th ed.), p. 727.
[39] V. O. Key, *Public Opinion and American Democracy* (New York: Alfred A. Knopf, 1961), p. 124.
[40] Wilson and Banfield, *op. cit.,* p. 876. Footnote 5 in the same article conveniently summarizes research supporting this proposition.
[41] Eugene S. Uyeki, "Patterns of Voting in a Metropolitan Area: 1938–1962," *Urban Affairs Quarterly,* I (June, 1966), pp. 65–77.

TABLE 4 CORRELATIONS BETWEEN MIDDLE CLASS CHARACTERISTICS
AND OUTPUTS IN REFORMED AND UNREFORMED CITIES

Correlations of	Government Type			Election Type		Constituency Type	
	Mayor-Council	Man-ager	Com-mis-sion	Parti-san	Non-parti-san	Ward	At-Large
Taxes with							
Median income	− .13	− .24	− .19	.03	− .19	− .17	− .22
White collar	− .23	− .12	− .62	− .21	− .33	− .30	− .32
Median education	− .36	− .22	− .08	− .45	− .24	− .48	− .18
Expenditures with							
Median income	− .19	− .32	− .43	− .04	− .32	−−23	− .34
White collar	− .24	− .23	− .58	− .18	− .39	− .32	− .35
Median education	− .32	− .36	− .26	− .36	− .38	− .44	− .32

expenditures from the national government. Therefore, we hypothesized
that:

2*a*. The more middle class the city, measured by income, education,
and occupation, the higher the municipal taxes and expenditures.

In line with our general concern of testing the impact of political structures
on municipal policies, we also hypothesized that:

2*b*. Unreformed cities reflect this relationship more strongly than reformed cities.

With respect to hypothesis 2*a*, the data in Table 4 on three middle
class indicators are unambiguous and indicate a strong rejection of the
hypothesis. However we measure social class, whether by income, education, or occupation, class measures are negatively related to public taxes
and expenditures.

It is possible, however, that income does not have a linear, but rather
a curvilinear relationship with municipal outputs. Banfield and Wilson
argue that "In the city, it is useful to think in terms of three income
groups—low, middle, and high. Surprising as it may seem to Marxists,
the conflict is generally between an alliance of low-income and high-income
groups on one side and the middle-income groups on the other."[42] If

[42] Banfield and Wilson, *op. cit.*, p. 35.

the relationship between income and expenditure is curvilinear, then we should expect to find that proportions of both low- and high-income groups are positively correlated with outputs. Our data, however, lend no support to this notion of a "pro-expenditure" alliance. Rather, the proportion of the population with incomes below $3000 is positively correlated with expenditures in all city types (although the relationships are small) and the proportion of the population in the above $10,000 bracket is negatively correlated with expenditures. Summing the two measures and correlating the combined measure with outputs produced no correlation greater than .15 and the relationships were as likely to be negative as positive. Tests for non-linearity also suggested that no such coalition exists in the cities in our analysis.

To be sure, aggregate data analysis using whole cities as units of analysis is no substitute for systematic survey data on middle-class attitudes, but it is apparent that cities with larger middle class population have lower, not higher expenditures. As we emphasized earlier, the "ethos theory" deals with attitudes and the behavior of individuals, while our data deal with cities and their behavior. The coalition suggested by Banfield and Wilson, however, is not discernible at this level of aggregation in these cities.

Hypothesis 2b is not consistently borne out by the data. In fact, the relationships between middle class variables and outputs are, if anything, stronger in the reformed than in the unreformed cities. One would not want to make too much out of the data, but a large body of literature on city politics, which we discuss below, suggests that reformed institutions maximize the power of the middle class.

We originally assumed that the proportion of owner-occupied dwelling units constituted another measure of middle class composition, but it soon became apparent that it was only weakly related to income, occupation, and education measures. Nevertheless, it emerged as the strongest single predictor of both expenditure and taxation policy in our cities. We hythesized that:

3a. Owner-occupancy and outputs are negatively correlated, and
3b. Unreformed cities reflect this relationship more strongly than reformed cities.

Hypothesis 3a is consistently borne out in the data presented in Table 5. These relationships were only slightly attenuated when we controlled for income, education, and occupation. No doubt self-interest (perhaps "private-regardingness") on the part of the home owner, whose property is intimately related to the tax structure of most local governments, may account for part of this relationship. Moreover, home ownership is corre-

TABLE 5   CORRELATIONS BETWEEN OWNER OCCUPANCY AND GOVERNMENT
OUTPUTS IN REFORMED AND UNREFORMED CITIES

Correlations of Owner Occupancy with:	Government Type			Election Type		Constituency Type	
	Mayor-Council	Manager	Commission	Partisan	Nonpartisan	Ward	At-Large
Taxes	− .57	− .31	− .73	− .64	− .45	− .56	− .48
Expenditures	− .51	− .23	− .62	− .62	− .40	− .50	− .40

lated (almost by definition) with lower urban population density. High density, bringing together all manner of men into the classic urban mosaic, may be itself correlated with factors which produce demands for higher expenditures—slums, increased needs for fire and police protection, and so on.

In confirmation of hypothesis 3*a*, the unmistakable pattern is for unreformed cities to reflect these negative relationships more strongly than the manager, nonpartisan and at-large cities, although commission cities show their usual remarkably high correlations.

Homogeneity Variables and Public Policies

Dawson and Robinson, in their analysis of state welfare expenditures, found strong positive relationships between the ethnicity of a state's population and the level of its welfare expenditures.[43] If this is symptomatic of a generalized association of ethnic and religious minorities with higher expenditures, we might find support for the hypothesis that:

4*a*. The larger the proportion of religious and ethnic minorities in the population, the higher the city's taxes and expenditures.

And, if our general hypothesis about the impact of political institutions is correct, then:

4*b*. Unreformed cities reflect this relationship more strongly than reformed cities.

The correlations between ethnicity, religious heterogeneity, and outputs (see Table 6) are, with one exception, positive, as predicted by hypothesis

[43] Richard E. Dawson and James A. Robinson, "The Politics of Welfare," in Herbert Jacob and Kenneth Vines, eds., *Politics in the American States* (Boston: Little, Brown and Co., 1965), pp. 398–401.

TABLE 6   CORRELATIONS BETWEEN ETHNICITY AND RELIGIOUS
HETEROGENEITY AND OUTPUTS IN REFORMED AND
UNREFORMED CITIES

Correlations of	Government Type			Election Type		Constituency Type	
	Mayor-Coun-cil	Man-ager	Com-mis-sion	Parti-san	Non-parti-san	Ward	At-Large
Taxes with							
Ethnicity	.49	.26	.57	.61	.43	.56	.40
Private school attendance	.38	.15	.37	.33	.37	.41	.25
Expenditures with							
Ethnicity	.36	.02	.21	.48	.21	.44	.13
Private school attendance	.34	−.01	.07	.25	.24	.40	.05

4*a*. These associations may reflect the substantial participation by ethnic groups in municipal politics long after the tide of immigration has been reduced to a trickle.[44] The relatively intense politicization of ethnic groups at the local level,[45] the appeals to nationality groups through "ticket balancing" and other means, and the resultant higher turnout of ethnic groups than other lower status groups,[46] may produce an influence on city government far out of proportion to their number.

We found when we related all twelve of our independent variables to outputs in various city types that the associations were much weaker in cities we have labeled reformed. The correlations for ethnicity and religious homogeneity show a generally similar pattern, with commission cities exhibiting their usual erratic behavior. The data, then, show fairly clear support for hypothesis 4*b*.

The third variable of our homogeneity indicators—per cent of population nonwhite—had almost no relationship to variation in outputs, regardless of city type. We found the same weak correlations for the poverty income variable, which was, of course, strongly related to the racial vari-

[44] Raymond Wolfinger, "The Development and Persistence of Ethnic Voting," *American Political Science Review*, LIX (December, 1965), pp. 896–908.
[45] Robert E. Lane, *Political Life* (Glencoe, Ill.: The Free Press, 1959), pp. 236–243.
[46] *Ibid.*

able. An easy explanation suggests that this is a consequence of the political impotence of Negroes and the poor, but one should be cautious in inferring a lack of power from the lack of a statistical association.

We have dealt in this section with factors which are positively and negatively related to spending patterns in American cities. While social class variables are associated negatively with outputs, two measures of homogeneity—private school attendance and ethnicity—are positively related to higher taxes and spending. Examining the strengths of these correlations in cities with differing forms, we found some support for our general hypothesis about the political consequences of institutions, especially for the homogeneity variables and the home ownership variable. Interestingly, however, this was not the case with class variables.

REFORMISM AS A CONTINUOUS VARIABLE

The central thrust of our argument has been that reformed governments differ from their unreformed counterparts in their responsiveness to socioeconomic cleavages in the population. Logically, if the presence of one feature of the "good government" syndrome had the impact of reducing responsiveness, the introduction of additional reformed institutions should have an additive effect and further reduce the impact of cleavages on decision-making. We therefore decided to treat "reformism" as a continuous variable for analytic purposes and hypothesized that:

5. The higher the level of reformism in a city, the lower its responsiveness to socio-economic cleavages in the population.

We utilized a simple four-point index to test this hypothesis, ranging from the "least reformed" to the "most reformed." The sample cities were categorized as follows:

1. Cities with none of the reformed institutions (i.e., the government is mayor-council, elections are partisan and constituencies are wards).
2. Cities with any one of the reformed institutions.
3. Cities with two of the reformed institutions.
4. Cities with three reformed institutions (i.e., the government is either manager or commission, elections are nonpartisan and constituencies are at-large).

We cannot overemphasize the crudity of this index as an operationalization of the complex and abstract concept of "reformism." Nonetheless, we think some of the relationships we found are strongly suggestive that reformism may in reality be a continuous variable.

TABLE 7   CORRELATIONS BETWEEN SELECTED INDEPENDENT VARIABLES
AND OUTPUT VARIABLES BY FOUR CATEGORIES OF REFORMISM

Correlations of	Reform Scores			
	1 (Least Reformed)	2	3	4 (Most Reformed)
Taxes with				
Ethnicity	.62	.41	.50	.34
Private school attendance	.40	.32	.28	.25
Owner-occupancy	−.70	−.39	−.54	−.44
Median education	−.55	−.27	−.32	−.13
Expenditures with				
Ethnicity	.51	.27	.41	.05
Private school attendance	.46	.23	.16	.08
Owner-occupancy	−.67	−.30	−.54	−.38
Median education	−.49	−.19	−.38	−.37

To test this hypothesis, we took four variables which had moderate to strong correlations with our dependent variables and computed simple correlations in each reform category. If our hypothesis is correct, the strength of the correlations in Table 7 should decrease regularly with an increase in reform scores. While there are some clear exceptions to the predicted pattern of relationships, there is some fairly consistent support for the hypothesis. Even when the decrease in the strength of the correlations is irregular, there is a clear difference between cities which we have labeled "most reformed" and "least reformed."

Again, we would not want to attach too much importance to the results of this rough-and-ready index. But the patterns support our previous argument about the impact of reformism: the more reformed the city, the less responsive it is to socio-economic cleavages in its political decision-making.

A CAUSAL MODEL AND AN INTERPRETATION

A Causal Model

The implicit, or at times explicit, causal model in much of the research on municipal reformism has been a simple one: socio-economic cleavages

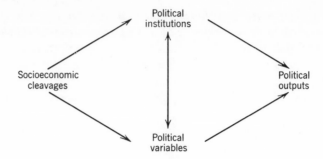

FIGURE 3   *A hypothesized causal model.*

cause the adoption of particular political forms. A more sophisticated model would include political institutions as one of the factors which produce a given output structure in city politics. We hypothesize that a causal model would include four classes of variables: socio-economic cleavages, political variables (including party registration, structure of party systems, patterns of aggregation, strength of interest groups, voter turnout, etc.), political institutions, (form of government, type of elections, and type of constituencies), and political outputs. Figure 3 depicts one possible causal model.

This study has of necessity been limited to exploring the linkages between socio-economic cleavages, political institutions, and political outputs. We found that political institutions "filter" the process of converting inputs into outputs. Some structures, particularly partisan elections, ward constituencies, mayor-council governments, and commission governments, operate to maximize the impact of cleavage indicators on public policies. We conclude by discussing some of the reasons that different structures have varying impacts on the conversion process.

An Interpretation

Three principal conclusions may be derived from this analysis.

1. Cities with reformed and unreformed institutions are not markedly different in terms of demographic variables. Indeed, some variables, like income, ran counter to the popular hypothesis that reformed cities are havens of the middle class. Our data lent some support to the notion that reformed cities were more homogeneous in their ethnic and religious populations. Still, it is apparent that reformed cities are by no means free from the impact of these cleavages.

2. The more important difference between the two kinds of cities is in their behavior, rather than their demography. Using multiple correlation

coefficients, we were able to predict municipal outputs more exactly in unreformed than in reformed cities. The translation of social conflicts into public policy and the responsiveness of political systems to class, racial, and religious cleavages differ markedly with the kind of political structure. Thus, political institutions seem to play an important role in the political process—a role substantially independent of a city's demography.

3. Our analysis has also demonstrated that reformism may be viewed as a continuous variable and that the political structures of the reform syndrome have an additive effect: the greater the reformism, the lower the responsiveness.

Through these political institutions, the goal of the reformers has been substantially fulfilled, for nonpartisan elections, at-large constituencies, and manager governments are associated with a lessened responsiveness of cities to the enduring conflicts of political life. Or, as Stone, Price and Stone argued in their study of changes produced by the adoption of manager governments, the council after the reform "tended to think more of the community as a whole and less of factional interests in making their decisions."[47]

The responsiveness of a political institution to political conflicts should not be confused with the "responsibility" of a political system as the latter term is used in the great debate over the relative "responsibility" of party systems.[48] In fact, the responsiveness of political forms to social cleavages may stand in sharp contrast to "responsible government" in the British model. Presumably, in American cities, partisan elections, ward constituencies, and mayor-council governments maximize minority rather than majority representation, assuring greater access to decision-makers than the reformed, bureaucratized, and "de-politicized" administrations.

Partisan electoral systems, when combined with ward representation, increase the access of two kinds of minority groups: those which are residentially segregated and which may, as a consequence of the electoral system, demand and obtain preferential consideration from their councilmen; and groups which constitute identifiable voting blocs to which parties and politicians may be beholden in the next election. The introduction

---

[47] Harold Stone, Don K. Price and Kathryn Stone, *City Manager Government in the United States* (Chicago: Public Administration Service, 1940), p. 238.
[48] The standard argument for party responsibility is found in the works of E. E. Schattschneider, esp., *Party Government* (New York: Farrar and Rinehart, 1942) and in the report of the Committee on Political Parties of the American Political Science Association, *Toward a More Responsible Two-Party System* (New York: Rinehart, 1950).

of at-large, nonpartisan elections has at least five consequences for these groups. First, they remove an important cue-giving agency—the party—from the electoral scene, leaving the voter to make decisions less on the policy commitments (however vague) of the party, and more on irrelevancies such as ethnic identification and name familiarity.[49] Second, by removing the party from the ballot, the reforms eliminate the principal agency of interest aggregation from the political system; hence, interests are articulated less clearly and are aggregated either by some other agency or not at all. Moreover, nonpartisanship has the effect of reducing the turnout in local elections by working class groups,[50] leaving officeholders freer from retaliation by these groups at the polls. Fourth, nonpartisanship may also serve to decrease the salience of "private-regarding" demands by increasing the relative political power of "public-regarding" agencies like the local press.[51] And when nonpartisanship is combined with election at-large, the impact of residentially segregated groups, or groups which obtain their strength from voting as blocs in municipal elections is further reduced.[52] For these reasons, it is clear that political reforms may have a significant impact in minimizing the role which social conflicts play in decision-making. By muting the demands of private-regarding groups, the electoral institutions of reformed governments make public policy less responsive to the demands which arise out of social conflicts in the population.

The structure of the government may serve further to modify the strength of minority groups over public policy. It is significant in this respect to note that commission governments, where social cleavages have the greatest impact on policy choices, are the most decentralized of the three governmental types and that manager governments are relatively the most centralized.[53] From the point of view of the reformer, commission government is a failure and their number has declined markedly in recent years.[54] This greater decentralization of commission and of mayor-council governments permits a multiplicity of access points for groups wishing to influence

---

[49] See Pomper, *op. cit.,* and Freeman, *op. cit.*

[50] Robert Salisbury and Gordon Black, "Class and Party in Partisan and Nonpartisan Elections: The Case of Des Moines," *American Political Science Review,* LVII (September, 1963), pp. 584–592.

[51] One newspaperman said of nonpartisan politics that "You can't tell the players without a scorecard, and we sell the scorecards." Banfield and Wilson, *op. cit.,* p. 157.

[52] Oliver P. Williams and Charles Adrian, *Four Cities* (Philadelphia: University of Pennsylvania Press, 1963), pp. 56–57.

[53] Alford and Scoble, *op. cit.,* p. 84.

[54] In our view, the failure of the commission government to achieve the intended reforms is more plausible as an explanation of its demise than its administrative unwieldiness—the conventional explanation.

decision-makers.[55] It may also increase the possibilities for collaboration between groups and a bureaucratic agency, a relationship which has characterized administrative patterns in the federal government. As a result of this decentralization, group strength in local governments may be maximized.

It is important in any analysis of reformism to distinguish between the factors which produce the *adoption* of reformed institutions and the *impact* of the new poltical forms once they have been established. We can offer from our data no conclusions about the origins of reformed structures, for it is obviously impossible to impute causation, using contemporary census data, to events which occurred decades ago. Once a city has institutionalized the reformers, ideals, however, a diffused attitude structure may be less helpful in explaining the city's public policy than the characteristics of the institutions themselves. With the introduction of these reforms, a new political pattern may emerge in which disputes are settled outside the political system, or in which they may be settled by the crowd at the civic club at the periphery of the system.[56] If they do enter the political process, an impersonal, "non-political" bureaucracy may take less account of the conflicting interests and pay more attention to the "correct" decision from the point of view of the municipal planner.

These conclusions are generally consistent with the ethos theory developed by Banfield and Wilson. If one of the components of the middle class reformer's ideal was "to seek the good of the community as a whole" and to minimize the impact of social cleavages on political decision-making, then their institutional reforms have served, by and large, to advance that goal.

[55] Williams and Adrian, *op. cit.,* pp. 30–31.
[56] Carol E. Thometz discusses the role of the "Civic Committee" in decision-making in Dallas. See *The Decision-Makers* (Dallas: Southern Methodist University Press, 1963).

PART III

Politics
and Education

ROBERT L. CRAIN

JAMES J. VANECKO

Elite Influence in School Desegregation

The question, "Who has influence?" is central to the study of local community government. In the last decade the question has often been asked in a more restrictive form: "How much influence does the business elite—the owners, managers, and executives of important economic institutions in the community—have?" Asked in this way, the question has provoked much controversy that shows little sign of reaching a resolution. The disputes are partly conceptual; there have been many attempts to define influence, and much of the disagreement has resulted from writers using different meanings of the word or referring to different kinds of influence. In addition, there have been problems collecting the needed data, problems both in the quality of the data and in the difficulty of obtaining comparative material from a number of cities.

This paper attempts to answer the question, "Who has influence?" by studying a single decision—the decision made in response to the demands made upon the public school system by the civil rights movement in a group of eight large Northern and Western cities. Obviously we cannot produce a general theory of power on the basis of a single decision in such a small number of cities. However, these limited data raise some rather new questions; and the second half of this paper includes speculations, resulting from these questions, about the structure of power in these cities.

We would like to express our gratitude to Gerald A. McWorter and Eve Weinberg for reading this paper and making a number of quite helpful comments. Also we wish to thank Bonnie McKeon for her capable use of the editorial pen.

THE RESEARCH DESIGN

The data were collected by the National Opinion Research Center in a study financed by the U.S. Office of Education.[1] The eight cities were chosen from a population including all cities of over 100,000. The cities were chosen in pairs, matching being based upon population size, per cent nonwhite, and region of the country. The interview data were collected by two researchers who spent approximately a week in each city attempting to uncover the story of the desegregation issue from various informants and interviewing the school superintendent, at least those members of the school board who were most involved, and whatever civil rights leaders, political leaders or civic leaders were identified as involved in the issue. The interviews were in part informal, probing for the respondent's role in the issue and for his interpretation of what happened, and in part formal, using short-answer questionnaires to determine the respondent's background and attitudes. The eight cities were Baltimore, Buffalo, Newark, Pittsburgh, St. Louis, San Francisco, and two cities identified here only by the pseudonyms, Bay City and Lawndale.

Both the advantages and the limitations of the techniques used should be obvious. The principal advantage is that we have a comparative study that can utilize concepts at various levels of abstraction ranging from the status of the population in the city (as measured by census data) down to the amount of influence exerted on a particular aspect of a particular decision (as reported by our interviewers). The principal disadvantages are that we are limited to eight cities and that in many cases we must settle for impressionistic data where additional time spent in the community might produce more persuasive evidence.

The eight cities in our study vary greatly in their handling of the school integration issue. At one extreme we have Baltimore, which agreed over the course of a single summer to bus large numbers of Negro students into previously white schools, or Pittsburgh, which produced a provocative report entitled "The Quest for Racial Equality,"[2] which to the outside observer reads more as if it were written by the civil rights movement than

---

[1] This study was supported by the U.S. Office of Education as a pilot study of educational decision-making. The original reports of the study are contained in Robert L. Crain with Morton Inger, Gerald A. McWorter, and James J. Vanecko, *School Desegregation in the North* (Chicago: NORC Report No. 110A, 1966), and *School Desegregation in New Orleans* (Chicago: NORC Report No. 110B, 1966). These two reports will be combined under one cover and published as *The Politics of School Desegregation* by the Aldine Publishing Company during 1967.
[2] "The Quest for Racial Equality" (Pittsburgh: Annual Report of the Pittsburgh Board of Education, 1965).

by a school board. At the other extreme are cities that had refused, at the time of our interviews, to take the first step toward school integration or that, indeed, had refused even to recognize the validity of the principle of school integration. The variance among these eight cities is summarized by a rank ordering which we shall call "acquiescence."

Acquiescence can be thought of as the extent to which the school board acted to bring the civil rights movement closer to its goals, both "welfare" and "symbolic." Thus acquiescence must consist of two elements: actions taken to further integration or upgrade education for Negroes, and actions that recognize the value of racial equality and the legitimacy of the civil rights movement. Acquiescence can be defined for any particular period of time, but throughout this paper we will define it for the entire period from the first raising of the issue to the time of our interviews.[3] This rank ordering, like most of those to be presented, is subjective. In this case it was developed by first having the interviewers fill out a questionnaire summarizing the actions taken by the school system. Armed with these questionnaires, the staff met several times to clarify the definition of acquiescence and agree upon a rank ordering. We cannot demonstrate with "hard" numerical data that this is the correct rank ordering.[4]

DIRECT INFLUENCE

When we consider the differences in the way in which the issue was handled in each city that might explain the variations in acquiescence, we find, rather surprisingly, that direct influence exerted on decision-makers does not seem to play a role. Even the level of activity of the civil rights movement does not seem to be correlated with acquiescence; an acquiescent city could be one that had very many demonstrations or one that had very few. In both acquiescent and nonacquiescent cities, mayors and other political leaders tended to remain silent when the issue arose, and in most cases none of the political leaders made any real effort to influence the decision.[5]

Most important for our purposes, even when the school board members are obviously closely tied to or part of the business elite of the community,

[3] This time period does *not* include the 1965 controversy in Pittsburgh nor the integration plan adopted by Buffalo that year.
[4] See Crain et al., *The Politics of School Desegregation,* for a discussion of the ranking procedure used in this study.
[5] We are using the term "political leaders" to refer to those persons who make political activity a full-time or almost full-time undertaking and who are successful at this on the local level. Thus, the term applies mainly to elected government officials and officers of the political parties.

we know of no cases in which school board members went back to the business leadership to solicit their views on a particular issue. This rather surprising statement is based on several facts: First, when school board members and superintendents were asked who was important in resolving the issue of school desegregation or who was influential on the decision made, they did not report any influence from outside the board, except in one case—Newark. In Newark, negotiations between civil rights leaders and the school board were initiated and arbitrated by the mayor. For the other seven cities, the negative response was given not only to a general question concerning influence; the economic elite also failed to turn up in the series of questions about those from whom advice was sought, with whom the problem was discussed, to whom information was given, and who were involved, interested, concerned, or worried about the outcome. A second piece of information confirming the noninfluence of the elite was that when persons who were frequently mentioned in answer to a general "reputational" question about influence in community decision-making were themselves interviewed, they expressed little interest in and displayed little knowledge of the desegregation issue. Perhaps the most important corroborating evidence was the fact that in a number of cases the members of the economic elite who were informed about the decision were in disagreement with or dissatisfied with the way the problem was handled by the school board (this is also true of three of the mayors). In at least two cities members of the local economic elite publicly disagreed with the school board, to no avail. Thus the image of an economic elite reigning over important community decisions does not fit the school desegregation issue.

In addition to the negative evidence cited above we have more persuasive positive evidence to make the same point. It can be shown that the differences in levels of acquiescence can be attributed almost entirely to differences in the character of the school boards. This means that outside influence could not affect the decision very much, since there would be little or no variance left to explain.

SCHOOL BOARD CHARACTERISTICS AND
SCHOOL BOARD BEHAVIOR

Each school board member was given a short agree-disagree questionnaire, to tap his general attitudes toward economics, civil rights, civil liberties, and other issues.[6] The school board in each city was assigned

[6] The questions included in the racial liberalism scale are fully reported in Crain et al., *op. cit.*

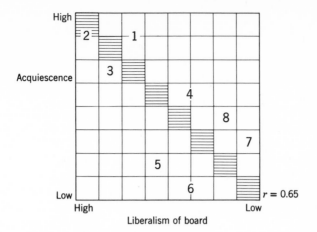

FIGURE 1 *Liberalism and acquiescence.* 1. *Pittsburgh;* 2. *Baltimore;* 3. *St. Louis (elected);* 4. *San Francisco;* 5. *Lawndale (elected);* 6. *Bay City (elected);* 7. *Buffalo;* 8. *Newark.*

the median value on the four-item civil rights scale administered to its members. The school boards vary considerably in this dimension, which we shall call simply "liberalism." When we plot liberalism of the school board against the acquiescence of the school system, we find the plot shown in Figure 1.

Two boards of the eight regularly have contested elections for school board membership. The remaining six are either appointed boards or boards where elections are uncontested. The two cities with contested school board elections score at the bottom on our acquiescence scale. In one of these two cities, our interviewers argue persuasively that the same board members serving in an appointive office would have been much more acquiescent.[7] In the other city, the mayor has violently disagreed with the way in which the school system has handled the civil rights issue, and it seems very likely that if this school board were appointed by the mayor or selected by a slating committee, the board might be quite different in its behavior.

For the remaining six boards, the story is very simple. There is a near-perfect rank-order correlation; the more liberal the school board, the

[7] The argument to show that this board would be more acquiescent contends that the strong ideological differences marking this board are the result of appealing to the different constituencies that must be served. If the same members served on an appointed board, they would be much freer to compromise their ideological differences and thus accede to some of the demands of the civil rights leaders.

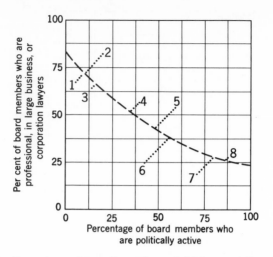

FIGURE 2   *Percentage of board members of high occupational status and percentage of members active in politics. 1. Pittsburgh; 2. Baltimore; 3. St. Louis (elected); 4. San Francisco; 5. Lawndale (elected); 6. Bay City (elected); 7. Buffalo; 8. Newark.*

more likely it is to meet the demands of the civil rights movement. This is true, despite the fact that the attitude questions used in the questionnaire had nothing to do with school integration and seemed to measure a stable attitude toward race relations. One rather surprising conclusion to be drawn from this is that the school board members react to the issue of school desegregation independently of the political forces and constraints inherent in their office. Personal characteristics of the school board members that are not related to the specific issue at hand but are in some way linked to their background prior to school board membership and to their recruitment to the school board determine the outcome of the issue. The winner can be known before the players step onto the field. This means that influences brought to bear on school board members, once involved in the issue, are not strong enough to overcome their personal values. A further conclusion, which is not a central concern of this paper, is that school policy in this issue is decided by the board and not by the superintendent, as the conventional wisdom has it.[8]

Why are some school boards more liberal than others? In Figure 2 we examine two characteristics of school board members aggregated for

[8] It is conceivable that the board members could have been socialized to these sets of attitudes through their experiences on the board. But since the indicators used had nothing to do with schools or education, this seems very unlikely.

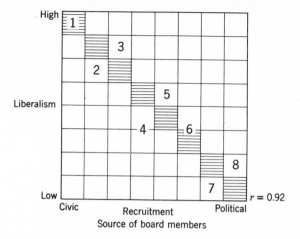

FIGURE 3   *Source of recruitment of school board members and board liberalism.*
1. *Pittsburgh;* 2. *Baltimore;* 3. *St. Louis (elected);* 4. *San Francisco;* 5. *Lawndale*
(elected); 6. *Bay City (elected);* 7. *Buffalo;* 8. *Newark.*

each of the eight boards. We see that there is a fairly strong negative
correlation between the number of school board members who have high
occupational status and the number of school board members who have
been directly active in partisan politics. At one extreme we have a school
board made up almost entirely of businessmen, corporation lawyers, and
other elites; at the other extreme we have a school board made up almost
entirely of persons previously active in party politics. To put it simply,
boards can be divided along a main dimension reflecting the extent to
which they have been "reformed."[9] Figure 2 contains a slightly curved
regression line, curved in order to take into account various threshold
and saturation effects. Using perpendiculars dropped to this line we can
rank the eight boards on the degree to which they recruit their school
board members from the two opposing camps of "civic" and "political"
candidates. The ranking is indicated by the numbers shown on the plot.
We can now ask what the relationship is between this and the liberalism
of the school board.

This relationship, as shown in Figure 3, presents a rather awesome
0.92 Spearman rank-order correlation. This table, therefore, represents
impressive support for the point of view of the so-called "pluralist"
school—that the persons having the authority to make the decisions are

[9] "Reform boards" refers merely to the style of recruitment, the types of persons
recruited, and the reasons for recruitment. The unreformed board is one to which
appointment is direct payment for political support.

in fact making them. In none of these communities can members of the school board be considered as the "power elite" of the city, certainly not the reputational elite.[10] In order to explain why one city chooses to acquiesce to the civil rights movement and another does not, we do not need to concern ourselves with the behavior of the civil rights movement, with local political competition, with the attitudes and behavior of the leading businessmen of the community, or even with the actions of the mayor. Rather we need only to find out whether the school system has a "reform" or a "political" school board, since this would tell us whether the board members are liberal or conservative, and that in turn would tell us whether the demands of the civil rights movement would be met or not.

INDIRECT INFLUENCE

Several writers have considered the kinds of indirect influence that might be important in local community decision-making. Peter Clark has discussed the ways in which leaders in positions of authority might anticipate the wishes of other elites and act in accordance with them, even though no direct influence has been exerted.[11] Robert Alford[12] and others[13] have discussed the fact that influence exerted at one point in time has the effect of establishing values that persist over long periods. Closely related to this is the way in which, over time, the community determines the kinds of persons who will reach positions of authority. Kent Jennings[14] has noted in his restudy of Atlanta that of the three decisions he analyzed, the civic elite was most active in the nomination of a candidate to replace Mayor Hartsfield. It seems obvious that if the civic elite can select the kind of mayor it wants, it will not need to be involved in the day-to-day decisions that he makes. This is essentially the case with the school system.

To test the hypothesis that a powerful business elite can influence recruitment to the school board, the cities were ranked according to the

[10] It is interesting to note that in only two cities did any member of the school board turn up in the answers to the reputational question.

[11] Peter B. Clark, "Civic Leadership: The Symbols of Legitimacy" (paper presented at the Annual Meeting of the American Political Science Association, September, 1960).

[12] Robert Alford, *Party and Society* (Chicago: Rand McNally, 1963).

[13] Peter Bachrach and Morton S. Baratz, "Two Faces of Power," *American Political Science Review,* LV (December, 1961); Delbert C. Miller and William Form, *Industry, Labor, and Community* (New York: Harper and Brothers, 1960).

[14] M. Kent Jennings, *Community Influentials* (New York: The Free Press of Glencoe, 1964).

influence of the elite in civic affairs independent of their activity regarding schools. The ranking is based on the report of informants in these cities, the report of the respondents in the cities, and the observations of the research staff. The questions on which the rank is based are of two types. First, in each city a group of five decisions that were considered the most important decisions of the past few years by our informants were used as probes in interviews to find the persons who had been influential. Second, the more general question—who is *reputed* to have influence—was asked, and this was followed by series of questions about the degree to which those named acted in unity, whether their influence was centralized in any organization such as the Greater Baltimore Committee or the Allegheny Conference in Pittsburgh,[15] or whether the men named were also influential in the decisions mentioned earlier. There were, of course, also questions concerning these men's influence in the school desegregation issue and the recruitment of school board members, but this factor did not have a strong effect on the overall ranking. When we compare the rank orders of the degree of reform in the school board and the influence of the economic elite in civic and political affairs we find a near-perfect correlation. (There is one reversal between adjoining cities, and one city ranked eighth on one scale is tied between sixth and seventh on the other.)

Thus, the fact that members of the economic elite are influential in the city in general seems to affect the kind of school board that is recruited. The logic behind this relationship is simple enough. If the economic notables are influential, then they are likely to be influential in school affairs. Their most likely point of intervention seems to be the recruitment of school board members, either to finance campaigns in a system that elects its school board or to confer status on the board in cities where members are appointed. Finally, the economic elite is likely to serve as a veto group with respect to the appointment or election of purely political members of the board; thus, influence wielded by an economic elite should prevent a reformed system from "backsliding."

Therefore, while we did not discover any direct influence of the economic elite in school desegregation, we did find considerable evidence of indirect influence through its role in the recruitment of the school boards. Unfortunately, only three of the school boards in this study were elected and five were appointed. Of the three elected boards, two showed

[15] Examples of such organizations are the Greater Baltimore Committee, the Allegheny Conference in Pittsburgh, and Civic Progress in St. Louis. Each of these organizations is composed of the top officers of the largest corporations in their respective cities. Each has had specific pet projects—Charles Center Urban Renewal Project in Baltimore, downtown renewal in Pittsburgh, the arch in St. Louis—but all of them also have been involved in a wide range of civic and political activity.

evidence of financial backing from various members of the economic and business elite—St. Louis and Lawndale. These were boards that had highly centralized internal patterns of influence. In St. Louis, all of the board members were part of a slate and owed most of their financial backing to the slate. In Lawndale, the self-perpetuating system of replacement on the school board assured centralized control. In the other system with an elected board, Bay City, there seems to have been no participation of the economic elite in the school board elections. (In fact the influence of the economic elite does not seem to extend to any local elections in Bay City.) Campaigns for school board seats do not cost a great deal in Bay City, and incumbents have the advantage of a full-time paid assistant who can serve as campaign manager. In the cities with appointed boards, there was no evidence uncovered of actual control of appointments by the economic elite, but we did find in Pittsburgh, Baltimore, and San Francisco that members of the economic elite were consulted about likely candidates. In some cases the mayor or the appointing agency gathered a large list of such candidates and seemed to have comparative freedom in choosing from this list. In other cases the more influential members of the board acted to locate potential recruits. In Newark and Buffalo, the appointments were purely rewards for political activity, and the business elite had almost nothing to say about them.

If the degree of reform in school appointments is related to general elite influence in the community, it follows that there should be a close correlation between this and other aspects of reform in the city. There seems to be some evidence for this; for example, the three school systems with the most "political" boards are in cities that have seen repeated charges of corruption leveled against the government.

THE CORRELATES OF REFORM

The simplest and most obvious hypothesis to explain why one city has reformed and another has not is to attribute this to differences in the population. It is commonly said that the classic political machine has its basis in the immigrant working-class and ethnically-identified population. Certainly the three cities with the most political boards qualify as having elements of the traditional political machine. When we examine the socioeconomic characteristics of the population in these cities, however, we see somewhat surprisingly that they are not particularly low status cities. In Table 1 the eight cities are assigned an average ranking based on each of six different socioeconomic characteristics. (In general, each of the individual variables correlates quite highly with the average ranking.

TABLE 1   SOCIOECONOMIC STATUS OF THE SCHOOL BOARD AND SELECTED
INDICATORS OF THE SOCIOECONOMIC STATUS OF THE
POPULATION OF THE COMMUNITY

Sample Cities	"Reform" School Board[a]	Annual Median Income	Median School Years Completed	Per Cent Earning Less than $5,000 Annually	Per Cent Earning More than $10,000 Annually	Per Cent with Some College	Per Cent in White Collar Jobs	Average of Rank
San Francisco	4	$6,717	12.0	31%	22%	24%	43%	1.0
Lawndale	5	6,303	11.4	35	17	19	37	2.0
Bay City	6	5,447	11.2	38	13	15	36	4.1
Buffalo	7	5,713	9.6	39	14	15	31	4.25
Baltimore	2	5,659	8.9	41	16	11	34	5.0
Pittsburgh	1	5,605	10.0	40	11	9	35	5.5
Newark	8	5,454	9.0	44	12	9	26	6.75
St. Louis	3	5,355	8.8	45	11	10	30	7.4

[a] Ranking is from 1 (most civic) to 8 (most political).

Of the six variables, annual median income is the poorest correlate with the average rank.) The three most thoroughly reformed cities are in the lower middle and bottom of the rank ordering; the three cities with the most machine-like political characteristics fall slightly above them. Apparently the overall socioeconomic status of the community is not related to the style of government. In fact this is not completely true; as we shall see later, the high status community both encourages and discourages elite influence over the schools.

We shall return to the relationship of population status to the community political style; for now it will suffice to say that there is no evidence here for the proposition that reform of school board appointments can be attributed to increased pressure from a more middle-class population.

THE INDUSTRIAL COMPLEXITY OF THE CITY

Another model that we believed to be somewhat more appropriate for the study of school board improvement is one of conflict—not between the professional politicians and the public but between the two most powerful elite groups in the city—the political party and the "civic leaders." By civic leaders we mean here not simply economic notables but rather the persons who have used their status and special skills to achieve leadership roles in civic affairs—service on the citizens' committees and in the fund-raising campaigns that Rossi has called "nondestructive potlatches." This is not necessarily a visible conflict; indeed there are many people whom we interviewed who would not accept this definition of the situation. We think that this conflict is simply a continuation of the pressures that divided these groups many years ago when the industrial cities in the North developed professional politicians who could use ethnic and class conflict as a resource to compete with Yankee money.[16] If the conflict is now somewhat less visible we believe it is because it has taken the form of establishing the norms governing the kinds of appointment that should be made. If the school board is appointed, the professional politicians "win," if the community can see that it is "only fair" to reward faithful politicians with seats on the school board. The civic elite "wins" one round when the mayor decides that the minority party should be represented; the battle is won when even bipartisan political appointments are considered taboo. This doesn't mean that the civic elite will be consulted on all appointments; it may mean only that the most prestigious members of the school board are consulted about possible candidates to fill a vacancy, but whatever the actual mechanism, the point is that in reformed cities it is expected that persons with personal status and skill will sit on the school board.

TABLE 2 SPEARMAN RANK-ORDER CORRELATIONS

		R
Per cent of labor force in manufacturing: city	x Influence of economic elite	.07
Per cent of labor force in manufacturing: SMSA	x Influence of economic elite	.19
Average size of manufacturing firms: city	x Influence of economic elite	.49
Average size of manufacturing firms: SMSA	x Influence of economic elite	.33
Combined ranking: industrialization	x Influence of economic elite	.17

From this point of view, the question is, "Why does the civic elite have more power in some cities than in others?" Here the work of Mills[17] and Fowler[18] suggests that cities with an economy dominated by a small

[16] Robert A. Dahl, *Who Governs?* (New Haven: Yale University Press, 1961).
[17] C. Wright Mills and Melvin Ulmer, *Small Business and Civic Welfare: Report on the Smaller War Plants Corporation to the Special Committee to Study Problems of American Small Business* (Washington, D.C.: Senate Document No. 135, 79th Congress, 2nd Session, 1946).
[18] Irving Fowler, *Local Industrial Structures, Economic Power, and Community Welfare* (Totowa, New Jersey: Bedminster Press, 1964).

bloc of manufacturing firms will have the most influential economic elite. (They disagree about whether such a domination is beneficial or not, but they agree that dominance results from a unification of manufacturing.) Table 2 presents rank-order correlations for the eight cities here studied among four operational variables suggested by Mills and Fowler. The highest correlation is between the average size of the manufacturing firms in the cities and the influence of the economic elite, 0.49. Given the size of this sample of cities this is not very high.

Since the correlations are in the direction predicted by Mills and Fowler, the best that can be said is that we have not yet presented conclusive evidence to support their point of view.

SUBURBANIZATION OF ELITES

The Mills-Fowler argument presupposes that the power of the business elite rests on their institutional connections, that elites participate on behalf of their corporations and use their connection with their corporations as their justification. But there is another point of view. Individual members of the civic elite may participate largely as individuals, and their particular corporation affiliation may not be of overriding importance. This point of view seems to have some support from our data. Thus, the man responsible for reform of the school board in St. Louis is a small businessman who himself has trivial economic power; he does have ties to other elites with more important economic resources, but there is no indication that he depends upon them for advice or that he defers to their presumably greater influence. In the area of schools he is the most influential member of the civic elite, primarily because he has the greatest energy, political skill, and educational experience.

As we look for variables that might explain the differential impact of the civic elite in these cities, we find one that is interesting and seems to support the conception of the participation of the elite. Our respondents mentioned the problem of recruiting executives who have moved to the suburbs. In Figure 4, we plot the "school board reform" variable against the relative suburbanization of the high-income families in the city. The relative suburbanization of high-income families is in this case measured by taking the proportion of families making over $25,000 per year who live outside the central city and dividing it by the proportion of the total population that lives outside the central city. Thus, for the city at the far left of the figure, this ratio is 0.8, meaning that the high-income families are slightly *more* likely to live in the central city than are other people. For the other seven cities, the relationship is above 1. When we examine the plot, we see that suburbanization is a fairly good predictor of reform. The correlation is 0.75, with two cities out of order. Note how difficult

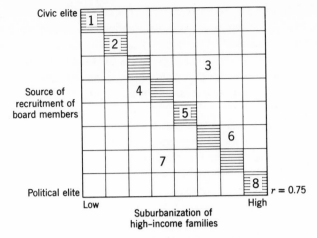

FIGURE 4   *Suburbanization of elites and school board reform. 1. Pittsburgh; 2. Baltimore; 3. St. Louis (elected); 4. San Francisco; 5. Lawndale (elected); 6. Bay City (elected); 7. Buffalo; 8. Newark.*

it is to reconcile this correlation with conventional theories of the "power elite." The whole idea of power structure implies a rigid set of relationships between persons in some sort of hierarchical structure, which again suggests that the heads of the largest corporations are in some sense the board of directors of the local civic elite.

Consistent with our work is George Sternlieb's finding that executives are more likely to be active in the civic affairs of the city in which they live than in which they work, whether that be central city or surburb.[19] This supports our finding that suburbanization tends to weaken the influence of the elites, but we cannot reconcile this with the power structure model, which implies either that civic activity on the part of businessmen is designed to at least indirectly benefit their business or is in some other way a function of their corporate ties, which would not be affected by suburbanization.

SPECULATION: THE CIVIC ELITE AS A "CLASS"

It may be that the civic elite can be described more accurately not as a structure of power but as a collection of individuals, each of whom has some resources and some contacts with other elites, who participate

[19] George Sternlieb, "Is Business Abandoning the Big City?" *Harvard Business Review* (January–February, 1961).

in community affairs as individuals but who constitute a diffuse "class" in that they have a common set of values. Much has been written about the withdrawal of business elites from local politics, but little has been said about their reentry into decision-making. Certainly the American city of the 1960's seems much more dependent upon the elite, who are serving on school boards, urban renewal commissions, Urban League boards, and so forth, than it was three decades ago. This return of the businessman to the city is not a return to the patterns of the nineteenth century. The business elite, it seems to us, has accepted the notion of a bifurcation between itself and the political professionals. Holding high status in the business community is worth little in an election campaign. In addition, the growth of national corporations selling to national markets, the shutoff in population growth of the central city, and the reform of city purchasing practices mean that few members of the business elite will reap any direct personal benefit from participation in politics. These two factors, we think, have led to the development of cooperation, rather than competition, among the elite and to the growth of a common ideology and agreement on goals which now permit the civic elite to behave as a class. The key elements in this set of goals are (1) a commitment to general economic development; (2) a commitment to reform; (3) a commitment to the public welfare, meaning both "amenities"[20] and charity on both the individual and governmental levels; and (4) a commitment to "keeping the peace" in the community.

These four goals—peace, prosperity, charity, and reform—constitute a common denominator around which the civic elite can agree. If the businessman moves beyond this framework, he may find that he has "become controversial." But within these limitations, he can expect the other members of the elite to give their endorsement to his actions. Within this framework, we think that the businessman participates not on behalf of his company but as an individual. The participation differs in degree but not in kind from the participation of his wife in PTA work or in the League of Women Voters. He participates because the work is entertaining, because it brings him prestige, and because of his desire to serve others. But beyond that, his participation furthers his class interests; he is helping to change the city into the kind of community which the members of his class, the civic elite, want. We do not need to postulate the existence of a power structure; the civic elite can remain merely a loose association of men who meet in the downtown clubs. If a man is invited to serve on the board of the Urban League he knows that his luncheon companions will generally approve. If he uses this position to begin some

[20] For a definition of "amenities," see Oliver P. Williams and Charles R. Adrian, *Four Cities* (Philadelphia: University of Pennsylvania Press, 1963).

program of action, he will have the tacit support of the other members of the elite, unless of course he commits some blunder or wanders outside agreed-upon goals. In fact, his participation may quickly make him known as the "specialist" in this area, the man to see for advice. By participating, the businessman enhances his status in the eyes of his colleagues and forms a common bond with them which gives him opportunities for increased interaction.

All of this makes the negative relationship between suburbanization and elite control of the school board more plausible. If in fact the civic elite is only a loose association of men who meet at lunch and on committees, then the conversation around the luncheon table will be heavily influenced by whether these are city dwellers who want to talk about city problems or suburbanites who let the conversation stray to other questions. In addition, many activities originate from our place of residence not from our place of work. Contributions to political parties, school activities, charities, residential conservation programs—all are examples of activities that might result from having our doorbell rung at home rather than at the office. Finally, and most important, the suburbanite has less information about or commitment to the central city. Another implication of this model is that the factor a member of the civic elite has that makes him valuable in civic affairs is probably not control over economic resources; rather, it is his personal skill, personal wealth, and willingness to work, coupled with his general high prestige and his contacts with other elites.

By saying that elites participate as a class rather than as a hierarchical power structure, we are suggesting that reform does not come about as the result of some backroom conversation between the political boss and the eldest member of the X family. Rather, we are suggesting a gradual and continuous grinding away by individual elites, through their participation (more or less as individuals) at many points in governmental systems and their continued gradual and individual influence upon political nominations.

It should of course be noted that this is no zero-sum game in which a victory for the civic elite implies a loss for the professional politician. Many professional politicians, even those heading very traditional machines, have decided that "good government is good politics." The civic elite may demand reform and appointment of "good men"; but they also provide useful services in the form of a great deal of time, energy, and skill. In addition, they supply support and some degree of legitimacy for the mayor, who can escape the responsibility for school affairs by claiming to have appointed the best possible people. This may explain, in part, why reform has been so successful in most American cities.

THE CONFLICT BETWEEN ELITES AND MASSES

The possibility of a negotiated truce between the professional politician and the members of the civic elite implies two things: first, that the professional politicians have the power to involve members of the elite in the decision-making process, and second, that these members are willing to settle for less than a complete take-over of all the political offices in the city. If these men were political candidates, there would be no grounds for compromise between them and the professional politicians. With these two considerations in mind let us look again at Figure 2 and Table 1. Three cities fall into the class of having reformed school boards and low-status populations: Pittsburgh, Baltimore, and St. Louis. In all three cases strong political parties are present. And while the cities differ in this regard, all three have ample room for the politician of working-class origins. The fourth city with a low-status population is Newark. Like the other three it has strong political parties, but its highly suburbanized elite has had only limited effectiveness in influencing public policy. In Newark, the school board seats are clearly in the party's hands. The other four cities are the four with high-status populations and much weaker political parties. Bay City and Lawndale are thoroughly nonpartisan, San Francisco's parties are very weak, and Buffalo's parties are badly splintered. One result of this is that in these cities it is very difficult to describe the typical school board member; some members are elected because of their political credentials, others because of their civic credentials, still others because of credentials too mysterious for us to understand. In contrast, the four strong-party systems tend, once we make allowance for the fact that every board has at least token ethnic and PTA representation, to have a consistent recruitment process that is clearly reflected in the kind of school board member recruited.

It seems quite plausible that the higher education, income, and occupational status of the citizens in these four cities has resulted over the years in either a weakening of existing political parties or at least in the prevention of the development of strong political parties. If we now reclassify these eight cities according to this second consideration, we can place the cities into the fourfold table shown in Figure 5. At the upper right-hand corner we have the highly organized cities with strong political parties that consistently recruit nonpolitical board members. These are cities with low-status populations but a civic elite that is not suburbanized. At the upper left we have cities that have an inconsistent pattern of recruitment: school boards containing both high-status persons and political activists. These are the cities where the elite is present but its influence is countered

Recruitment process is:

	Inconsistent	Consistent
Civic	Middle–class cities ——————— San Francisco Lawndale*	"Balance of power" cities ——————— Pittsburgh Baltimore St. Louis
Political	Working–class cities ——————— Buffalo Bay City*	"Machine" cities ——————— Newark

Recruitment source

*Pseudonyms

FIGURE 5   *A typology of political structure and school board recruitment.* 1. *Pittsburgh;* 2. *Baltimore;* 3. *St. Louis (elected);* 4. *San Francisco;* 5. *Lawndale (elected);* 6. *Bay City (elected);* 7. *Buffalo;* 8. *Newark.*

by the presence of a strong middle-class vote. At the lower left we have two school boards that select persons, again without a consistent pattern, who are not particularly high-status, whether they are in politics or out of it. Both of these are relatively high-status cities but with weak political parties and inactive civic elites; while we have called them working-class cities they might also be described as lower–middle-class cities. Finally, in the fourth cell, we have the cities that are strictly political in their identification of board members. The one city falling into this cell has a low-status population and a suburbanized elite.

To review, the two main dimensions that account for school board composition are, first, the overall influence of the civic elite, which seems to have its roots in (among other things) the extent to which the elites have stayed in the city, and, second, the strength of political parties. The civic elite cannot win "control" of the school board as easily if political parties are weak.

This leads to some intriguing speculations. First, we would argue on the basis of what we have said so far that the civic elite can wield influence in reforming a city and in the kinds of local decisions made only if it is not permitted direct participation in local politics. The civic elite can either participate as a nonpolitical class, or it can participate as a group of politicized individuals. Therefore, we reform a city by forcing the elite out of politics.

The other finding, which is a corollary of this, is that the civic elite can wield influence only in working-class cities. From the elites' point of view, the ideal city is one which has an upper class and a working class but no middle class. The same pattern holds across the bottom of the table. Buffalo and Bay City have higher-status populations than Newark; the result is that they have weak political parties but are still not middle-class enough to replace working-class politics with middle-class politics. The result is a high level of political disorganization and a very low level of acquiescence to the civil rights movement.[21]

How can it be that the existence of civic elite influence depends upon the existence of strong political parties, when we have earlier argued that it is the very conflict between these two institutions that is basic to our understanding of these cities? One factor is that the strong political party serves as a barrier to prevent participation on the part of individual elites. Thereby, it encourages them to act in concert, as a class. But in addition, and probably more important, the existence of strong political parties serves as a barrier to prevent the mass opinion of the middle classes from having a great influence in elections or policy formation because the party is able to coalesce the disparate interests of large segments of both the middle class and the working and lower classes. Thus, the heads of government and political professionals are free to accept the advice and seek the consent of the civic elite in policy formation and public appointments without having to worry about the general public's opinion. Therefore, strong political parties free candidates from strong ideological commitments and permit compromise between the goals of these candidates—achievement and maintenance of office—and the goals of the civic elite—peace, prosperity, charity, reform. Finally, the existence of strong political parties means that politics is organized, which in turn means that if the civic elite wishes to exert influence (particularly in small and gradual bits) it can do so because there is a central point (the party leadership) upon which this influence can be brought to bear. If, on the other hand, politics is completely disorganized, then the exertion of pressure at one point in time tends not to have any particularly strong cumulative effect on the mayors who will succeed the incumbent in office. This is even more true if the school board is elected. Bay City, for example, which has very weak political parties, has seen the school board repeatedly reformed, but each reform dies out again. In cities with stronger

---

[21] Edward C. Banfield and James Q. Wilson present a similar argument in *City Politics* (Cambridge, Massachusetts: Harvard University Press, 1963).

political parties, the school board tradition is expressed in the kind of men who are slated for office.

To complete our discussion at this abstract level, let us turn to the two census characteristics that we believe to be most directly related to these two dimensions of the political style of the school system. One is the suburbanization of the elite, the other is the general socioeconomic status of the city. In Figure 6, we have presented these two variables together and identified the eight cities again. Notice that the cities in the upper right, which have a low socioeconomic status but an elite living in the central city, are the ones that should be most highly reformed and therefore the most acquiescent to the demands made by the civil rights movement. Conversely, at the lower left we have cities which are high status but

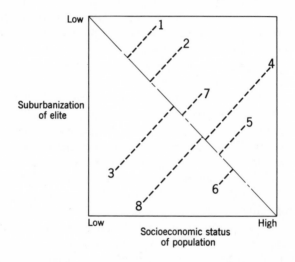

Identification Number	Name	Acquiescence Rank
1	Pittsburgh	1.5
2	Baltimore	1.5
3	St. Louis	3
6	Buffalo	6
5	Newark	5
4	San Francisco	4
7	Lawndale	7
8	Bay City	8

Spearman rank order correlation = 0.90

FIGURE 6   Acquiescence and two census variables.

whose elites have moved to the suburbs. It is these cities that we anticipate will be least acquiescent to the civil rights movement. If we draw a line from the lower right-hand corner to the upper left-hand corner and place the eight cities on this line, we find a rank-order correlation of 0.89 between the combined effect of these two census variables and the outcome of the school segregation decision.

CONCLUSION

In analyzing the school desegregation issue, we have presented a model of the city that revolves around three actors: the civic elite, the professional politicians, and the mass of voters. We have argued that the professional politicians play two roles. As actors, they are the conservative or intolerant school board members who refuse to desegregate schools and precipitate extreme racial conflict.[22] And as a political party, they block the masses from direct participation. The result of this is to permit the businessman who wishes to participate in nonpolitical community affairs to wield more influence, to be eligible for school board membership; thus, the strong party permits recruitment of those men who are most able to produce a plan to peacefully desegregate schools.

The most important distinction made by this model is between three kinds of participation on the part of elites: political participation, individual "civic" participation, and participation as representatives of their economic institutions. The two so-called middle-class cities of our typology are cities where elites participate directly in partisan politics. In these cases it is possible to appoint a semiprofessional politician who, to the casual observer, may have all the attributes of a "good" appointment; in fact, however, this person is not a member of the "civic elite" since he has not demonstrated his skill in nonpolitical civic leadership. Furthermore, the appointment of these sorts of persons injects the elite into politics and thus encourages personal rivalries and ideological disagreements. Both middle-class cities have elites that are divided by these conflicts, and it is our hypothesis that the failure of the system to insulate them from day-to-day political affairs has contributed to this. The other distinction, between participation as an individual and participation as a repre-

---

[22] One explanation of the apparent inability of politicians on school boards to negotiate with civil rights leaders is the type of politician recruited to the school board. Political appointments to the school board are normally rewards for activity during a campaign or campaigns, which usually means delivering a ward. These politicians are ward leaders and thus have no previous experience in arbitrating the conflicting interests within a party but have rather represented one of those particular interests.

sentative of a corporation, is more subtle but also important. The goals implied in these two kinds of behavior can be quite different and they are often in conflict. The traditional bifurcation of the political and civic elites was in part based upon an intense conflict of interest. Now with the advent of personal participation of the elite this conflict is moderated. The large corporations have less to gain from local politics and less chance of gaining it. In this light, the elites become nonrepresentative, that is, they represent no particular interest or group. This adds to the system, first, an arbitrator between the particular interests of various groups and, second, a voice for those not represented in the system, such as the poor, the Negroes, or the mobile middle class; in addition, it provides a spokesman for the shared values of the civic elite class—the values we have earlier referred to as peace, prosperity, charity, and reform.

It is of course not always possible to distinguish between institutional participation of the economic elite and personal participation. The executive who proposes local tax reform might be concerned both with more efficient and equitable government and with possible benefits for his own company. What can be clearly distinguished are the bases for the two types of participation. Institutional participation is based upon a shared interest of the economic institutions of the community. Personal participation is based upon the shared values of individuals. If members of the new elite do not represent the whole community, they certainly represent a larger segment of it than did their firm fifty years ago when it was busy buying streetcar franchises. Thus in community conflicts the representatives of leading corporations can take stands on public issues, such as fair housing laws and school desegregation, can work in cooperation with the government in the economic promotion of the city, can commit their own resources to the development of urban renewal plans, can advance their own education programs, and can even work to promote metropolitan government—all activities that have taken place in the reformed cities that we studied.

PART IV

Politics
and Law Enforcement

JOHN A. GARDINER

Police Enforcement
of Traffic Laws:
A Comparative Analysis

VARIATIONS IN TRAFFIC LAW ENFORCEMENT

Over the last few years, a small but growing number of studies have been made of the outcomes of urban political systems. *Comparative* urban studies, however, have been greatly hampered by difficulties in locating outcomes which can be measured; while it is easy to ascertain the presence or absence of urban renewal programs or the dollars spent on education, it is generally difficult to say whether City A's programs are "different

This study was originally delivered as a paper at the 1966 Annual Meeting of the American Political Science Association. The research on which this study is based was conducted while the author was a V. O. Key, Jr., Fellow of the Joint Center for Urban Studies of the Massachusetts Institute of Technology and Harvard University, and a dissertation fellow of the Woodrow Wilson National Fellowship Foundation. The author also gratefully acknowledges research support from the University of Wisconsin Faculty Research Committee and the Russell Sage Foundation Program in Sociology and Law at the University of Wisconsin. Richard Collins collected and calculated most of the data involved in the study of 508 cities. Keith Billingsley provided computer programming advice. Computer time at Harvard and the University of Wisconsin was supported by the National Science Foundation. Valuable comments and suggestions were offered by James Q. Wilson, Herman Goldstein, Herbert Jacob, Donald McCrone, Kenneth Dolbeare, James W. Davis, Jr., and Rufus P. Browning.

from," to say nothing of "better than," City B's—we are seldom sure exactly what these figures "measure."[1] In the administration of criminal justice, general problems of "measuring" outputs are compounded by the uneven and frequently inaccurate collection of basic statistics,[2] and even where the police and courts are reporting data accurately, it is still very difficult to determine whether, in several cities, "similar" defendants are given "similar" treatment following the commission of "similar" offenses.[3] Working backward in the process of criminal justice, attempts to discover whether the "same" people are brought into the process in the first place, i.e., to discover whether the police in City A apprehend and charge *all* persons who have violated a law, while the police in City B either charge only certain classes of violators (Negroes, juveniles, recidivists, etc.) or else ignore a law completely, run into the problem of ascertaining how often the laws have been violated. Does the fact that B's police have arrested fewer persons than A's reflect a difference in policy or simply the commission of fewer crimes? In most cases, we cannot fill this data gap by assuming equality (e.g., by assuming that every city sees a similar number of each type of offense per 1,000 population) and then concluding that the variations in the number of arrests are a measure of variations in police policy. FBI figures on offenses known to the police, while of

[1] See James Q. Wilson, "Problems in the Study of Urban Politics," a paper delivered at a conference in Bloomington, Indiana, November 5–7, 1964; and H. Douglas Price, "Comparative Analysis in State and Local Politics," a paper delivered at the 1963 Annual Meeting of the American Political Science Association, September 6, 1963. Works comparing urban policy outcomes include Robert R. Alford and Harry M. Scoble, "Political and Socioeconomic Characteristics of American Cities," *1965 Municipal Yearbook* (Chicago: International City Managers' Association, 1965); Oliver P. Williams and Charles R. Adrian, *Four Cities: A Study in Comparative Policy Making* (Philadelphia: University of Pennsylvania Press, 1963); Oliver P. Williams, Harold Herman, Charles S. Liebman, and Thomas R. Dye, *Suburban Differences and Metropolitan Policies* (Philadelphia: University of Pennsylvania Press, 1965); Louis H. Masotti and Don R. Bowen, "Communities and Budgets: The Sociology of Municipal Expenditures," *Urban Affairs Quarterly,* I (December, 1965), pp. 39–58.

[2] See Ronald H. Beattie, "Problems of Criminal Statistics in the United States," *Journal of Criminal Law, Criminology, and Police Science,* XLVI (July–August, 1955), pp. 178–186; Institute of Public Administration, *Crime Records in Police Management,* (New York: Institute of Public Administration, 1952); Marvin E. Wolfgang, "Uniform Crime Reports: A Critical Appraisal," *University of Pennsylvania Law Review,* CXI (April, 1963), pp. 708–738; James Q. Wilson, "Crime in the Streets," *The Public Interest,* No. 5 (Fall, 1966), pp. 26–35. For an illustration of the results following a change in police crime reporting policies, see Eric Pace, "'True' City Tally of Crime Pushes Rates Up Sharply," *New York Times,* April 5, 1966.

[3] See Roger Hood, *Sentencing in Magistrates' Courts: A Study of Variations of Policy* (London: Stevens and Sons, 1962); Paul W. Tappan, *Crime, Justice, and Correction* (New York: McGraw-Hill, 1960), pp. 443–445; Edward Green, *Judicial Attitudes in Sentencing: A Study of the Factors Underlying the Sentencing Practice of the Criminal Court of Philadelphia* (London: MacMillan, 1961).

questionable accuracy,[4] indicate that there are substantial differences in the crime rates of different cities. In the area of *traffic* law violations (violations of laws regulating *moving* traffic—speeding, stop signs, red lights, etc.—but *not* parking), however, an assumption of equality might perhaps not be so tenuous. It seems somewhat more reasonable to assume that drivers in every city in the country have an equal propensity to speed or to coast through stop signs than to assume a constant rate of burglary or prostitution.

To test the proposition that traffic cops in different cities enforce the laws equally, letters were sent in 1965 to the chiefs of police of the 697 cities, towns, and townships with 1960 populations greater than 25,000. Five hundred and eight (73 per cent) returned information concerning 1964 ticketing and the size and organization of their forces.[5] To standardize these figures with some estimate of the volume of traffic in the city (and thus, by hypothesis, of the volume of traffic violations), the number of tickets written was divided by the number of motor vehicles registered in that city in 1964 and by the city's 1960 population. Since this study focuses upon the role of the police, the number of tickets written was also divided by the number of full-time policemen (excluding clerical and maintenance workers) on the police force. The distribution of responses on each of the three measures is presented in Tables 1, 2, and 3.[6]

Put in words, looking at only one of these measures, 173 police forces ticketed less than 50 persons per 1,000 population, while 36 forces ticketed more than 250 persons per 1,000. Figures from a few pairs of cities of similar size will dramatize the differences involved. The police in Boston and Dallas, two cities which in 1960 had populations of approximately 700,000 wrote 11,242 and 273,626 moving violation tickets respec-

---

[4] See sources cited in footnote 2, *supra*.

[5] To check the possibility that the chiefs would overstate their ticket figures, responses from chiefs in one state (Massachusetts) were compared with the actual lists of traffic tickets submitted to the state agency charged with driver license control. Of the 180 chiefs reporting (in this state, all municipalities larger than 5,000 were surveyed), almost none varied by more than 1 or 2 per cent from the official records; occasional *large* variations resulted from the inclusion of *parking* tickets with the moving violations figures. In the national study, any figures which differed markedly from those of cities of similar size were verified through a second letter to the chief to see if parking tickets had been included.

[6] Pearsonian product-moment correlations between the three measures are:

Tickets/1,000 motor vehicles—Tickets/1,000 population $r = .89$
Tickets/1,000 motor vehicles—Tickets/Policeman   .82
Tickets/1,000 population—Tickets/Policeman   .85

Information on the number of tickets and the number of policemen were obtained from each police department; information on the number of motor vehicles registered in each city was obtained from state motor vehicle departments or safety organizations; population figures are from the 1960 Census.

tively—a 24-fold difference! Niagara Falls, New York, and Wichita Falls, Texas, both slightly over 100,000 in 1960, produced 1,245 and 10,211 tickets in 1964. Police in Springfield, Massachusetts, wrote 14,720 tickets while their counterparts in Grand Rapids, Michigan (also with a popula-

TABLE 1   1964 TICKETS PER 1,000 REGISTERED
               MOTOR VEHICLES

Range	N	Per Cent
0– 49	74	14.6
50– 99	98	19.3
100–149	74	14.6
150–199	72	14.2
200–249	51	10.0
250–299	46	9.1
300–349	29	5.7
350–399	18	3.5
400–449	13	2.6
450 and over	27	5.3
Unknown	6	1.2
Total	508	100.1[a] $\bar{X} = 185, s = 144$

[a] Totals for Tables 1 and 2 do not add to 100 per cent because of rounding.

TABLE 2   1964 TICKETS PER 1,000 (1960) POPULATION

Range	N	Per Cent
0– 49	173	34.1
50– 99	127	25.0
100–149	83	16.3
150–199	61	12.0
200–249	28	5.5
250–299	19	3.7
300 and over	17	3.3
Total	508	99.9 $\bar{X} = 105, s = 87$

tion of 175,000) were writing 36,727. Finally, in Cambridge and Somerville, Massachusetts, two adjacent cities with populations of 100,000, the police wrote 5,457 and 750 tickets respectively.

In the absence of data on the *actual* number of traffic violations committed in each city during the year 1964, these three sets of figures—tickets per 1,000 motor vehicles, tickets per 1,000 population, and tickets per policeman—were offered as gross measures of the intensity of traffic law enforcement in these cities. It can easily be argued, however, that there is little relationship between the ultimate data sought (actual violations) and the substitute bases (motor vehicles, population, policemen) used; drivers in one city may be more law-abiding than those in another, traffic congestion and street layout may make speeding more difficult in some cities than in others, some cities may depend more on the automobile as a means of transportation, and so forth. The "tickets per 1,000 motor vehicles" figure ignores traffic passing through the city; the "tickets per

TABLE 3   1964 TICKETS PER POLICEMAN

Range	N	Per Cent
0– 24	125	24.6
25– 49	114	22.4
50– 74	68	13.4
75– 99	65	12.8
100–124	64	12.6
125–149	23	4.5
150–174	25	4.9
175–199	11	2.2
200 and over	12	2.4
Unknown	1	0.2
Total	508	100.0 $\bar{X} = 70, s = 58$

1,000 population" figure also ignores population changes between 1960 and 1964. All of these objections are valid. Yet it is hard to believe that the probable discrepancy between the number of actual violations and the bases used can explain all of the variation between the cities—the fact that the Dallas police ticketed 2,334 per cent more motorists than the Boston police, the fact that the Cambridge police ticketed seven times as many motorists as the police in neighboring Somerville, or the fact that 173 police forces ticketed less than 50 persons per 1,000 population while 36 forces ticketed more than 250 persons per 1,000.

TRAFFIC LAW ENFORCEMENT IN THE POLICE DEPARTMENT

If these ticketing figures in fact reflect significant variations, several interpretations are possible. First, it will be argued that the ticketing variations are based on variations among departments concerning the conduct expected of policemen; in some police forces, the men are actively encouraged to write traffic tickets, while in other departments, other aspects of police work are stressed at the expense of traffic. Second, on the basis of statistical correlations between the ticketing figures and the census data, it will be argued that the intensity of traffic law enforcement varies to a certain extent with the social structure of the community, possibly because of differences in citizen expectations of the police, but more likely because of the impact of rapid changes in the composition of the community on police attitudes toward the public.

Before analyzing *variations* in attitudes toward traffic work within different police departments, it might be useful to begin with a discussion of certain aspects of traffic work which are common to all policemen.[7] It has frequently been stated that since traffic duties entail contact with citizens who are otherwise law-abiding, policemen will often try to minimize citizen hostility by ignoring minor violations or by letting drivers off with a warning. Antagonism generated among motorists (the police fear) may cut off future cooperation in apprehending criminals or support for police requests for increases in staff or salaries.[8] Against this factor must be placed certain aspects of traffic enforcement work which can make it a comparatively pleasant assignment within the police department. Unlike the regular patrolman who must walk a beat and investigate suspicious occurrences, the traffic officer faces a minimum of physical labor or danger. He is also less likely to have frequent contact with a sergeant or other superior officer and usually does not have to report in to headquarters as regularly as a beat man. As one Massachusetts patrolman put it, "Out here on the road, nobody bothers me. I don't have to call in or take orders. So long as I hand in my tickets at the end of the day, I can pretty much do what I please."

This comment illustrates a major facet of traffic work in the eyes of many policemen—it is easy for the traffic man to prove to his superiors that he is "on the job." While the general assignment officer lacks a

[7] The data contained in this and the following paragraphs (where not otherwise footnoted) are based upon personal observation of and (unstructured) interviews with police chiefs and officers in approximately thirty cities and towns in Massachusetts. As yet, there are no studies of police attitudes toward their work based upon formal, structured interviews with policemen in different cities, although a study associated with the President's Commission on Law Enforcement and Administration of Justice is presently working in this area.

[8] The impact of traffic enforcement work on police-public relations is noted in Jerome H. Skolnick, *Justice Without Trial: Law Enforcement in Democratic Society* (New York: Wiley, 1966), pp. 54–56; T. C. Willett, *Criminal on the Road: A Study of Motoring Offenses and Those Who Commit Them* (London: Tavistock, 1964), Chap. 4; and O. W. Wilson, "Police Authority in a Free Society," *Journal of Criminal Law, Criminology, and Police Science*, LIV (June, 1964), pp. 175–177.

demonstrable "product" for his hours of patrolling, investigating, and questioning, the traffic officer can keep his superiors "off his back" simply by demonstrating through the number of tickets he hands in that he has been on the job. Where a traffic-oriented captain or chief expects a high rate of ticket-writing from his traffic men, they may be kept busy fulfilling these expectations, but they can still safeguard their independence.[9] Finally, it might be noted that traffic men usually take no *personal* interest in traffic offenders, although an abusive comment may lead them to write a ticket when they had planned only a warning.[10] In one set of cities studied by the author in which ticket fixing by the chief was common, a custom has developed by which the chief will always contact the arresting officer to see if the motorist was courteous; if he was abusive, it is understood that the officer can "veto" the proposed fix, and occasionally an officer will write on the ticket, "He was abusive, chief. Please don't fix this one."[11]

Just as traffic duties impose conflicting pressures upon individual policemen, they also mean several things to the police chief. As in other areas of law enforcement,[12] the chief is caught in the midst of conflicting social demands and expectations—reduce accidents but act "reasonably"; ticket as many violators as possible but keep the budget down. A policy of strict traffic enforcement can produce certain fringe benefits such as a

[9] In one Massachusetts city in which the chief demanded active ticketing on the part of his men, one officer met his "quota" easily after he located a personal "gold mine"—a stop sign so frequently ignored by motorists that he could produce as many tickets as he had time to write; standing on this corner, the officer managed to write 1,161 tickets during a four-month period in 1965—21 per cent of the tickets written in that period by the entire police force of one hundred men. Cf. Skolnick, *op. cit.*, p. 55, quoting a California traffic cop: "You learn to sniff out places where you can catch violators when you're running behind. Of course, the department gets to know that you hang around one place, and they sometimes try to repair the situation there, but a lot of the time it would be too expensive to fix up the engineering fault, so we keep making our norm."

[10] The impact of abusive conduct in causing the policeman to act more severely and with more personal involvement is noted by William A. Westley, *The Police: A Sociological Study of Law, Custom, and Morality* (unpublished Ph.D. dissertation, University of Chicago, Department of Sociology, 1951); Westley, "Violence and the Police," *American Journal of Sociology*, LIX (July, 1953), pp. 34–41; Westley, "Secrecy and the Police," *Social Forces*, XXXIV (March, 1956), pp. 254–257; Wayne R. LaFave, *Arrest: The Decision to Take a Suspect into Custody* (Boston: Little, Brown, 1965), p. 146; Michael Banton, *The Policeman in the Community* (London: Tavistock, 1964), p. 137; Skolnick, *op. cit.*, p. 146; and Joseph Goldstein, "Police Discretion Not to Invoke the Criminal Process: Low-Visibility Decisions in the Administration of Justice," *Yale Law Journal*, LXIX (March, 1960), pp. 543–594.

[11] John A. Gardiner, *Traffic Law Enforcement in Massachusetts* (unpublished Ph.D. dissertation, Department of Government, Harvard University, 1965).

[12] See James Q. Wilson, "The Police and their Problems: A Theory," *Public Policy*, XII (1963), pp. 189–216; Skolnick, *op. cit.*

legitimate method of harassing juveniles or a pretext for interrogating and searching suspicious persons.[13] On the other hand, a policy of leniency (e.g., in fixing tickets) can provide a form of "patronage" which can be exchanged for higher salaries or new equipment from local political leaders or reduced prices for merchandise or services from local tradesmen.[14] Finally, traffic duties enmesh the chief in a network of decisions involving resource allocation—faced with a statutory duty to enforce *all* laws, a limited amount of men and funds must be distributed among them. An extra patrolman can be assigned to gambling or murder—or traffic; traffic enforcement can be assigned to *all* men "when you're not doing anything else" or to a special traffic detail. Regardless of what attitudes the chief may have towards traffic enforcement on an *a priori* basis, they must ultimately be translated into man-hours and funds, resources which if used for traffic cannot be used for other police functions.[15]

These cross-pressures on the traffic officer and his chief—be courteous to citizens yet see that the laws are enforced; reduce accidents but keep costs down—are present in all cities. What accounts for the variations among cities? The basic difference among police departments seems to be the extent to which chiefs and other commanding officers expect a high rate of ticketing from their men and distribute incentives to achieve that end. While, as will be shown shortly, certain procedural differences related to court appearances can affect police attitudes toward traffic, the most important factor affecting a policeman's decision to cite or ignore traffic violators is the demand for tickets by superior officers. Where the incentive system[16] of the department is used to reward active ticket-writers,

[13] On the use of the traffic stop to interrogate suspects, see LaFave, *op. cit.,* p. 187; and Donald M. McIntyre, Lawrence P. Tiffany, and Daniel Rotenberg, *Detection of Crime* (Boston: Little, Brown, 1966).

[14] Gardiner, *op. cit.;* Banton, *op. cit.*

[15] Several police chiefs, particularly in smaller departments, attributed a *decline* in ticketing between 1963 and 1964 to an increase in other duties. "We had to pull men off traffic and put them on night shift to cut the crime rate." "The department had more criminal work and so took officers off patrol for investigation." "We had to spend more man-hours on other crimes." Unfortunately, it was impossible to tell whether these "increases" in crime were objectively correct or rather reflected changing perceptions by the chief of his several responsibilities.

On the general problem of resource allocation in police departments, see Herman Goldstein, "Police Discretion: The Ideal Versus the Real," *Public Administration Review,* XXIII (September, 1963), pp. 140–148; Frank J. Remington and Victor G. Rosenblum, "The Criminal Law and the Legislative Process," *University of Illinois Law Forum* (Winter, 1960), pp. 481–499; LaFave, *op. cit.* pp. 70–71 and Chap. 5.

[16] By incentive systems, I am referring, of course, to the set of inducements which the leaders of an organization offer to obtain cooperation of any desired nature from members of the organization. See Chester I. Barnard, *The Functions of the Executive* (Cambridge: Harvard University Press, 1938); Peter B. Clark and James

the police will respond with tickets; where department norms are neutral or openly hostile toward traffic enforcement, ticketing will be low and only serious offenders will be cited.

The way in which organizational norms (and thus the goals toward which an incentive system is directed) influence the ticketing policies of traffic patrolmen can be seen in illustrations from several cities in which police attitudes toward traffic work were suddenly and dramatically shifted. In 1958, one Daniel Brennan was named chief of a police department

---

Q. Wilson, "Incentive Systems: A Theory of Organizations," *Administrative Science Quarterly,* VI (September, 1961), pp. 129–166; Peter M. Blau, *The Dynamics of Bureaucracy* (Chicago: University of Chicago Press, 1955). While, in most police departments, the major *material* incentives affecting policemen (wages, hours, fringe benefits, etc.) are determined by state and local laws and ordinances, there are many ways in which the chief is relatively free to seek to influence the attitudes of his men. First, with varying degrees of independence, the chief can impose formal penalties, such as demotion in rank, loss of pay, etc., upon men who fail to meet required standards. Second, the chief can define the duties of his men (e.g., the number of blocks to be patrolled, the frequency with which a man must report in to headquarters, the length and frequency of coffee breaks, etc.) and assign men to different posts. A chief thus can transfer a patrolman from the day to the night shift, from a pleasant residential beat to a "tough" section of the city, or from the detective squad to the "drunk wagon." One specific way in which the chief can define duties so as to maximize the production of traffic tickets is by setting up a specialized traffic detail, rather than asking *all* men on the force to look for violations. One traffic expert has declared that the generalized policeman is incapable of good traffic work "because of lack of training, administrative skill, and competent, interested supervision." Franklin Kreml, "The Specialized Traffic Division," *Annals,* CCXCI (January, 1954), p. 71. After observing specialized and nonspecialized traffic men in a number of cities, it would seem that this statement is correct, but for the wrong reasons. While special education and training may be necessary for accident investigation and traffic engineering, a patrolman can be trained in ten minutes to judge the speed of a car and to decide whether a motorist has come to a "full stop." The reason why specialization of traffic enforcement will lead to more ticketing seems rather to be that when writing tickets is the *only* function of a policeman, the number of tickets written becomes the sole measure of his "work-product." Even without any active coercion from superiors or any personal sense of mission or ideological attachment to traffic work, a specialized traffic man will write tickets simply to prove he is "on the job" and thus to avoid criticism. As a third method of influencing work patterns, the chief is free to assign tasks that can produce extra income for his men. These include overtime work and "pay details," when an outside person pays the department to provide extra police protection, e.g., at a construction site, a wedding, or a department store sale. For either kind of work, the patrolman will receive pay in addition to his regular salary. Fourth, the chief can also influence the performance of his men through his training and retraining programs, whether operated through a formal school or simply through assigning new recruits to older officers for instruction. Finally, and probably most important, the chief can attempt to affect the conduct of his men through such informal mechanisms as commendation of exemplary conduct, keeping track of the public and private lives of his men, and so forth.

in a New England city. Unlike his predecessor, who had spent most of his police career as a detective and felt that nondetective aspects of police work were a waste of time, Brennan felt that his men should be active in *all* phases of law enforcement, especially including traffic law enforcement. To increase the number of tickets written, Brennan changed from a system of work assignments under which *all* men on the force were supposed to write tickets in addition to their specific duties to one in which three men were assigned, on a full-time basis, *only* to enforcement while others were given enforcement duties from time to time. These "specialists," while only four or five men in a force of 235, wrote over 60 per cent of the force's 5,500 tickets in 1964. Going beyond this formal process of specialization, Brennan sought to arouse greater interest in traffic enforcement by instituting a daily work report system by which *all* men in the department had to list how many investigations and arrests were made and how many traffic and parking tickets were written each day. When a man's reports indicated noticeably fewer "results" than were reported by other men with similar assignments, he was called in and questioned by the chief. Brennan also pushed ticketing by example (going out on his lunch hour and writing tickets himself) and inducement (officially commending high ticket producers and giving overtime enforcement work to those who "produced" tickets on their regular jobs). Despite frequent dislike of Brennan and the demands he made of them, the specialists preferred traffic work to walking a beat, and most regular men on the force preferred writing tickets to having Brennan "on their backs." In 1957, the last year of the former police chief, the entire force wrote only 480 tickets. In 1959, the force wrote 4,569 tickets, and in 1963, 7,381 tickets—a 14-fold increase over 1957!

The process by which Chief Brennan reoriented attitudes toward traffic work finds a parallel in a small midwestern suburb. The chief of this 28-man department gave this explanation as to why his men wrote 881 tickets in 1963 and 3,605 in 1964:

Since we are a small department . . . we felt that we could not afford to become specialists, but rather we had to become generalists. Hence we set about the task of making *all* officers traffic enforcement conscious. . . . Officers are never required to fill a quota where traffic tickets are concerned, but are simply ordered to enforce existing laws vigorously.

The Inspector's Office maintains a breakdown of daily records on all police personnel relative to their activities. These records, of course, almost immediately put the finger on the "goldbrick," or one who is not doing his job to full capacity. In this event, the officer in question is ordered into the Inspector's Office where comparative records are made available to him and efforts are made to determine reasons for the officer's lack of initiative. If the conference does not procure the desired results, further and more severe disciplinary action usually follows.[17]

It should also be noted that a police department's incentive system can be used to *reduce* the number of tickets written. In one New England city in which the 185-man police force wrote 502 tickets in 1964 (or 14 per 1,000 motor vehicles as compared with a national mean of 185), the chief explained, "We just want our men to use common sense; we only want to write tickets when the violation is beyond a reasonable doubt." In case an exceptionally obtuse patrolman in this city is in doubt whether to err in the direction of zeal or moderation, he can remember the attitude of a former chief of the department. This chief told the author of a patrolman who ticketed a motorist driving 50 m.p.h. along a busy two-lane street in the center of the city. Called by the irate motorist, the chief summoned the patrolman to his office. "Well, did you ask him *why* he was doing fifty? Did you find out if he was late for work? Was his wife sick? . . ."

In general, the rate of enforcement of traffic laws reflects the organizational norms of a police department—the extent to which superior officers expect and encourage particular policies regarding ticket-writing. Under certain circumstances, however, the rate of enforcement can be influenced by one factor *outside* the department, a court-established procedural difference regarding appearance in court. In some cities (126 of the 508), the arresting officer must go to court on the day on which the motorist has been ordered to appear, even though the motorist may already have paid his fine or decided to plead guilty (in which case the officer's testimony will be unnecessary). In these "first call" cities, an officer assigned to the day shift will simply be sitting in court rather than working; an afternoon or night shift officer, on the other hand, will be called into court when he is off duty. A night shift officer, for example, might work until 7 A.M., have to appear in court at 9:30, and not be home in bed until noon. In other cities, however, the arresting officer is only called into court after a motorist has pleaded not guilty, an event which occurs in only 10 to 15 per cent of all minor traffic cases. While no traffic officer will admit that he will not write out a ticket because of an obligation

[17] Letter to the author from a Missouri police chief, February 23, 1966 (emphasis in original). A further example of the effects upon ticketing of a change in the incentive system of a police department can be seen in data from Chicago. Following a series of scandals, the number of tickets written by the Chicago Police Department fell from 570,000 in 1959 to 355,000 in 1960. After traffic-conscious O. W. Wilson was appointed as Superintendent in 1960, traffic enforcement was required of both traffic and district police, and the totals rose to 681,000 tickets in 1962. See "Chicago Record Again Proves 'Three E' Power," *Chicago Traffic Safety Review* (January–February, 1963).

TABLE 4 1964 TICKETS PER POLICEMAN IN FIRST CALL AND NOT
GUILTY CITIES

Range	First Call N	First Call Per Cent	Not Guilty N	Not Guilty Per Cent	Total[a] N	Total[a] Per Cent
0– 24	46	36.5	74	21.0	125	24.6
25– 49	25	19.8	81	22.9	114	22.4
50– 74	18	14.3	41	11.6	68	13.4
75– 99	13	10.3	50	14.2	65	12.8
100–124	14	11.1	47	13.3	64	12.6
125–149	5	4.0	18	5.1	23	4.5
150–174	1	0.8	24	6.8	25	4.9
175–199	1	0.8	9	2.5	11	2.2
200 and over	3	2.4	8	2.3	12	2.4
Unknown[b]	—	—	1	0.3	1	0.2
	126	100.0	353	100.0	508	100.0
	$\bar{X} = 57, s = 55$		$\bar{X} = 74, s = 55$		$\bar{X} = 69, s = 55$	

[a] The Total column includes 29 cities whose chiefs did not indicate in their responses whether the local court required appearance by the officer on the first call or only after the defendant pleaded not guilty.
[b] The chief of one "not guilty" city did not disclose the number of men in his department.

to appear in court, many officers in the "first call" cities complain of hours wasted standing in court.[18] Table 4 shows the difference in ticketing rates between the "first call" and "not guilty" cities. Summarizing this data, the mean of the "first call" cities was 57 tickets per policeman; where the arresting officer could wait until the defendant had pleaded not guilty, the mean was 74 tickets per policeman.[19] When these figures

[18] The impact of court appearance rules on the arresting practices of various types of enforcement officers is noted in American Bar Foundation, *The Administration of Criminal Justice in the United States: Pilot Project Report,* Volume II (Chicago: American Bar Foundation, 1957), p. 288 (Detroit traffic police); James Q. Wilson, "The Police and the Delinquent in Two Cities," included in this volume (juvenile officers in two cities); and Steven V. Roberts, "Wasted Time Cut in Housing Court," *New York Times,* June 19, 1966 (New York building inspectors).
[19] Using the other measures of the intensity of ticketing, the comparable figures are:

	Tickets per 1,000 Vehicles	*Tickets per 1,000 Population*
All cities	$\bar{X} = 185, s = 144$ $(N = 502)$	$\bar{X} = 105, s = 87$ $(N = 508)$
First call cities	$\bar{X} = 164, s = 152$ $(N = 124)$	$\bar{X} = 87,\ s = 82$ $(N = 126)$
Not guilty cities	$\bar{X} = 191, s = 138$ $(N = 349)$	$\bar{X} = 110, s = 89$ $(N = 353)$

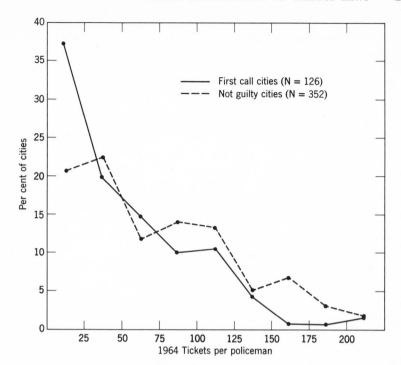

FIGURE 1   *Distribution of ticketing rates in first call and not guilty cities. The vertical axis represents the per cent of each type of city (first call or not guilty) in each range of ticketing. Thus, for example, 37 per cent of the first call cities and 21 per cent of the not guilty cities fell in the range of 0–24 tickets per policeman. The area under each line equals 100 per cent of the cities of that type.*

are plotted on a graph (see Figure 1), however, it can be seen that the difference in means is produced by the large number of "first call" cities in the lowest ticketing ranges. In the middle ranges (50 to 150 tickets per policeman), there is little difference between the two types of cities. In the higher ranges, the "not guilty" cities predominate, although the small number of cities makes analysis difficult.

One explanation for the differential impact of the court appearance requirement may be that when departmental incentives to write tickets are *not* present, the court appearance factor will be critical in the mind of the policeman; where he is ordered to write tickets, the distinction will be less important.[20] This possibility is supported by data from Massa-

[20] I am greatly indebted to my colleague Rufus Browning, Jr., for interpretive suggestions on this point.

chusetts police forces, where ticket-writing is seldom a major organizational goal: in contrast with the national mean (for 502 cities over 25,000) of 185 tickets per 1,000 motor vehicles, 118 "first call" cities and towns (over 5,000 population) in Massachusetts averaged only 38 tickets per 1,000 motor vehicles while the mean for 62 "not guilty" municipalities was 84 tickets. "Not guilty" police forces in Massachusetts ticket at a rate 121 per cent higher than that of the "first call" forces.[21]

TRAFFIC ENFORCEMENT IN THE COMMUNITY

The primary factor leading to a high or low rate of traffic enforcement, therefore, is the organizational norm or policy fostered by the chief and commanding officers, modified in some cities by the first call-not guilty court requirement. Next, it might be asked whether these organizational norms are associated in any way with the characteristics of the communities the police departments serve. (Since there are no "hard" data on department norms, the ticketing results of these norms must be used as a substitute for the norms in the following analysis.) If such correlations exist, two explanations might be offered—first, that police policies are determined by community expectations and are just as much "outputs" of the local political system as are revenue or expenditure policies. A second explanation of any correlation between traffic policies and community characteristics would be that the police are influenced not by community expectations but rather by the character of the "clientele" with whom they are dealing—that the police respond differently to different types of citizens.

The first proposition, that police enforcement policies would parallel variations in the community because of varying "community expectations" of the police, demands assumptions that people in the community care whether traffic laws are enforced strictly or leniently, that these popular attitudes vary according to social class or some other demographic grouping, and that the police know of and carry out these community-based policies. The possibility that members of a community *care* about police traffic enforcement policies might be bolstered by the "objective" importance of these policies. With the exception of the schoolteacher, the traffic policeman is probably the municipal employee with whom the average citizen has the greatest personal contact. In addition, police work accounts for a significant portion of the municipal budget—the mean per cent of general expenditures devoted to police work in the 508 cities studied was

[21] Gardiner, *op. cit.*

12 per cent.[22] (While there is no evidence that a "strict" traffic enforcement policy would be either more or less expensive than a "lenient" policy,[23] the size of the police budget might at least attract some public interest.) Finally, police enforcement policies would seem to be important to the community because of the commonly accepted view (to be questioned later in this paper) that "strict enforcement" will promote traffic safety and thus reduce insurance rates.[24]

Even if it is accurate to say that people in a community care what policies their police follow, and that traffic policies are "important," there are several reasons why we should *not* expect to find correlations between ticketing rates and demographic characteristics of the communities. First, in a nation so generally tied to the automobile, *all* communities may have similar attitudes toward traffic law enforcement—attitudes simultaneously desiring traffic safety and the avoidance of tickets. Early declarations that the traffic violation was primarily a "white collar" crime have been seriously questioned,[25] and while different traffic laws may hit different classes (e.g., lower-class drivers are more likely to violate safety equipment and insurance regulations while upper-class drivers are more frequently arrested for driving while under the influence of alcohol),[26] all classes may be equally (and equally ambiguously) interested in whether the local police will give them a ticket or save them from accidents.

More important than the probable similarity of attitudes toward traffic law enforcement in all cities in the weakening of the assumptions which could relate ticketing policies to community variations is the fact that traffic enforcement *policies* (as opposed to individual actions) tend to be completely unknown or irrelevant to the community. Most traffic violations do not involve a "victim" who would have a personal interest in police action. Furthermore, since most police officials try to foster a public

[22] Data on police expenditures are taken from the 1962 Census of Governments.
[23] The distribution of receipts from traffic fines varies from city to city; in some cities, a portion of these fines goes to the police department; in others to the city treasury. Unfortunately, it was impossible to ascertain facts on this matter for the cities contained in this study; thus it was impossible to determine whether police forces which "profited" directly from ticketing had higher rates. Even within one state (Massachusetts) in which *all* cities had a common method for disposing of traffic fines, there was great variation in the intensity of police traffic activity. See Gardiner, *op. cit.*
[24] See, e.g., O. W. Wilson, *Police Administration,* 2nd ed. (New York: McGraw-Hill, 1963), Chap. 17.
[25] H. Laurence Ross, "Traffic Law Violation—A Folk Crime," *Social Problems,* VIII (Winter, 1960–61), pp. 231–241, reports that the proportion of white collar workers arrested for traffic violations is higher than that involved in other types of crime, but also that they are represented among those arrested for this type of crime roughly in proportion to their numbers in the community (here, Evanston, Illinois).
[26] Willett, *op. cit.,* pp. 194–199.

impression that *all* violations will be punished, they seek to conceal any departmental policies which would lead to ignoring one class of crime (e.g., minor traffic violations), concentrating upon violations in one part of the city, or setting an official "tolerance" level (e.g., specifying that tickets will only be written when the motorist exceeds the speed limit by 15 m.p.h.).[27] Thus, even if they were interested in police ticketing policies, it would be almost impossible for citizens to discover them. Finally, the validity of these assumptions might be questioned on the basis of the argument by William Westley and others that the police, by the very nature of their function, are isolated from the rest of the community; this isolation, we might hypothesize, would vitiate the impact of the community (and of variations among communities) on the activites of the police.[28]

In the course of personal interviews with approximately thirty police chiefs in Massachusetts,[29] the following picture of the "political" or "community" context—the social pressures which the *chiefs* felt were brought to bear upon them—of traffic law enforcement emerged. Almost no one in a city or town *knows* whether the local police force is enforcing traffic laws *in general* in an active or passive manner or particularly *cares* which policy is followed. There is probably a community consensus, the chiefs feel, that *serious* violations (driving under the influence of alcohol, reckless driving, etc.) should be prosecuted by the police as vigorously as

[27] LaFave, *op. cit.,* pp. 153–157; Joseph Goldstein, *op. cit.;* Ross, *op. cit.;* Herman Goldstein, *op. cit.*

[28] See Westley, works cited in footnote 10, *supra;* James Q. Wilson, "Police Morale, Reform, and Citizen Respect: The Chicago Case," in David Bordua, ed., *The Police* (New York: Wiley, 1967), pp. 137–162. Recent comparative studies have indicated that while American policemen are more socially isolated than workers in other occupations, they are at least less isolated than their British counterparts. Structured interviews with 611 British and 313 American policemen indicate that whereas British police attribute their isolation to public hostility and suspicion, the American police (here studied in three middle-sized Illinois cities) attribute their isolation more to their peculiar working hours, the unique demands of their work, and public dislike for representatives of arbitrary authority. The British public views each policeman as a representative of the police establishment; American officers are more likely to be regarded as individuals. To the extent that the comparison is correct, American policemen might at least be more likely than their British counterparts to feel the attitudes of the community and to reflect variations among communities. See John P. Clark, "Isolation of the Police: A Comparison of the British and American Situations," *Journal of Criminal Law, Criminology, and Police Science,* LVI (September, 1965), pp. 307–319. The British data are contained in R. Morton-Williams, "Relations between the Police and the Public," printed as Appendix IV of Royal Commission on the Police, *Minutes of Evidence* (London: H.M.S.O., 1962). Michael Banton's comparative study of two Scottish and three American police departments reaches a similar conclusion (*op. cit.*).

[29] Gardiner, *op. cit.*

crimes of violence or crimes against property. On the other hand, there is probably a community feeling that traffic officers should display *some* degree of leniency; if one policeman suddenly becomes a "ticketing fool"—ticketing old ladies going one mile over the speed limit or the proverbial husband rushing his pregnant wife to the hospital—the community will expect the chief to "sit on him" or "bury" him in some post where he cannot antagonize the public. Between these two extremes, however, the chiefs interviewed gave no indication that their communities cared what enforcement policies they adopted.

The general pattern of community disinterest did not mean, however, that the chiefs were subject to *no* public pressure with regard to traffic law enforcement. Every police department is subjected to a barrage of calls from city councilmen, parents, and other citizens relating to *specific* demands—"Put an officer by the grade school." "Stop those kids from 'peeling rubber' by the drive-in." "Could you have your men arrest those speeders on Main Street?" And, in the other direction, "Can't you do something about that speeding ticket my son got yesterday? That was his first offense." "Chief, could you tell your men to lay off Main Street during the Christmas shopping season? They're hurting business." Note that all of these demands have a very *ad hoc* quality; once the caller sees a police car at the grade school or drive-in, he feels that he has "gotten action," and will neither know nor care whether the police are active at *other* schools or restaurants. To be sure, local newspapers regularly run safety editorials, and, in the larger cities, safety councils call for greater safety awareness on the part of drivers and enforcement activity on the part of police; the chiefs interviewed, however, expressed the opinion that these campaigns have little influence on drivers and none on them.[30] It might be added that officials of two safety organizations in Massachusetts agreed with this conclusion; most chiefs, they felt, ignored their demands for stricter enforcement. Feeling that police cooperation on safety education and engineering matters was preferable to a public showdown, the safety officials chose not to press the enforcement issue in the newspapers or before local officials.[31]

For all of these reasons, the hypothesis that police traffic enforcement policies will vary according to varying "community expectations" rests upon at best very questionable assumptions. To see whether it has any statistical support, however, the three measures of the intensity of traffic law enforcement were compared, using a Pearsonian *r* correlation program,

---

[30] *Ibid.*
[31] Interviews with officials of the Massachusetts Safety Council and the Western Massachusetts Safety Council, September, 1964, and February, 1965.

TABLE 5   PRODUCT-MOMENT CORRELATIONS BETWEEN TICKETING
AND POPULATION MOBILITY[33]

	Tickets per 1,000 Vehicles	Tickets per Policeman	Tickets per 1,000 Population
Per cent native of state	$r = -.38$	$r = -.44$	$r = -.48$
Per cent same house 1955–1960	$-.36$	$-.45$	$-.43$
Per cent new house since 1958	$.34$	$.42$	$.41$
Per cent different county, 1955–1960	$.25$	$.38$	$.34$

with a number of gross measures of the social characteristics of the 508 cities—1960 Census data on income, education, ethnicity, and housing.[32] On none of the measures of ticketing (tickets per 1,000 motor vehicles, tickets per policeman, and tickets per 1,000 population) were there correlations greater than $\pm.30$ with measures of income (median family income, per cent with incomes lower than $3,000 or higher than $10,000) or education (median school years completed, per cent with less than 5 or more than 12 years of schooling). The same was true for measures of ethnicity (per cent foreign born or with foreign or mixed parentage), race (per cent nonwhite), and housing (per cent unsound structures and median home value).

Moderate positive correlations appeared, however, when ticketing rates were compared with population mobility; cities with highly stable populations had generally lower rates of ticketing than cities with a high proportion of new residents. The correlations are presented in Table 5.

How do these statistical correlations affect the two hypotheses presented earlier—that police policies are the product of community expectations or that the police will respond differently to different types of "client" citizens? The lack of correlation between ticketing and demographic characteristics suggests either that "community expectations" do not vary according to these characteristics, or else that, for the reasons stated above,

[32] I am indebted to Professor Robert R. Alford of the University of Wisconsin Department of Sociology for the use of his card decks of 1960 Census data.
[33] To check for the possibility that some variable associated with mobility was producing these moderate correlations, partial correlations were obtained on one measure of ticketing. The partial correlation between tickets per policeman and per cent living in the same house in 1960 as in 1955 (simple correlation $= -.45$) was $-.38$ when controlled for median school years, $-.46$ when controlled for median family income, and $-.46$ when controlled for per cent in white-collar occupations. Mobility, therefore, had an independent role in producing correlations with ticketing.

there are no meaningful community expectations regarding police traffic law enforcement policies—the public neither knows nor cares much about these policies. The moderate correlations indicated between ticketing and population mobility, however, suggest that community expectations may vary with the length of residence of the population rather than with any static characteristics. Looking at the form of government of all American cities of more than 25,000 population, Alford and Scoble found that population mobility (here, the per cent living in a different county in 1960 than in 1955) was the best predictor of whether a city would have a council-manager rather than a mayor-council form of government.[34] Assuming that the mayor-council form "encourages or allows interest group representation" while the council-manager form "encourages efficient implementation of specified goals," Alford and Scoble hypothesize that population growth and mobility are "intervening variables, serving to loosen the social and political ties of persons to their community and rendering ineffective those characteristics of the population which would otherwise bring forth political demands."[35] In highly mobile cities, therefore, where political interests have not yet become organized, the goal of efficiency will predominate and be reflected in the city-manager form of government. The parallel hypothesis might be offered that in the absence of "political demands" to enforce laws with reference to persons, the police will be likely to enforce criminal laws strictly (with reference to rules); further, just as the stabilization of a population produces demands which can be satisfied more readily through a mayor-council form of government, so will stabilization produce demands for the "political" enforcement of criminal laws. Strict law enforcement might thus be expected in the newer, more mobile communities where countervailing political pressures have not yet developed.

The correlations between mobility and ticketing also lend support to the second, alternative hypothesis stated earlier, that policemen respond differently to different types of "clientele" populations with whom they must deal. Michael Banton, studying two Scottish and three American police forces, concluded that there was a direct relationship between the level of social integration (the level of consensus or agreement on fundamental values)[36] and the utility of informal sanctions against lawbreakers. Police in more integrated cities feel that informal controls will produce desired changes in conduct and thus are less likely to impose formal sanctions; where the agreement on fundamental values is lower, the police

---

[34] Alford and Scoble, *op. cit.* Mobility remained the best predictor of the form of local government even after partialling for region, occupation, population change, and city size.
[35] *Ibid.,* p. 96.
[36] Banton, *op. cit.,* p. 3.

may well feel that only formal sanctions will be effective.[37] An example of this can be found in the tendency, noted earlier, for police to issue tickets to abusive motorists; such conduct, the police say, indicates "disrespect for the law," and thus a probability that only formal measures (tickets) will be effective. Since there is a fairly strong ($-.48$) negative correlation between the stability of the population (here, the per cent native of the same state) and FBI "Part I" Crimes per 1,000 population, it would not be unreasonable to assume (although there are no hard data on this point) that policemen would expect better results from informal sanctions (e.g., oral warnings) in older, more settled communities than in communities with rapidly changing populations.[38]

Further testing of these hypotheses must await more extensive research into both the value systems of police departments and the process of interaction between police and local political leaders. A study of value systems may explain more fully the factors which influence the officer in deciding how to enforce laws. What is the relative importance of personal background and the socialization process involved in training programs? Are "strict" responses to one set of laws (e.g., traffic laws) carried over to other laws (e.g., gambling, prostitution, vagrancy, or narcotics)? Research on the latter question is necessary to learn how (and if) community expectations regarding the level of enforcement of criminal laws are (1) formulated, and (2) transmitted to the police. Are severity-leniency controversies handled in the same manner as other disputes over public policy or in some other fashion, because they are invisible, politically profitless, or "too hot to handle"? If normal political processes are *not* involved, which political forces are thereby benefitted or diminished in power? Finally, is it appropriate or even useful to think of local law enforcement within a "systems" framework of analysis?[39] (More broadly, does the local system of criminal justice acquire the characteristics of the local political system?)

This study of variations in police ticketing practices also raises several questions of public policy. First, there is the question whether police are the appropriate agency for the enforcement of traffic laws. While few empirical studies have been made on this point, there is some evidence

[37] *Ibid.*, pp. 136–137.

[38] The correlation between population mobility and crime rates might suggest that the police in more mobile communities might be working with a greater rate of traffic violations per 1,000 vehicles, per 1,000 population, per policeman, etc., and thus that the seeming differences noted earlier don't really exist—that the police forces are in fact responding similarly to varying rates of violation of traffic laws. When the relationship between ticketing (tickets per policeman) and mobility (per cent native of the same state) is controlled for crimes per 1,000 population, however, the correlation remains moderately high ($-.36$).

[39] David Easton, *The Political System* (New York: Alfred Knopf, 1953).

that police enforcement rates have *no* influence whatsoever on the rate of traffic violations or accidents.[40] Indeed, one political scientist, who dealt with traffic safety matters while a member of the Harriman administration in New York State, has argued that traffic violations should be handled not as crimes but rather as problems of public health, requiring the services of doctors, psychologists, and engineers rather than crime-trained police.[41] To the extent that the police conduct studied here was *not* based on safety principles (as when the ticketed violation was not hazardous or when the officer was simply trying to fill his quota), this reassignment of the traffic function would make sense. It should be pointed out, however, that police administrators feel that traffic control is vital to police work for *nonsafety* reasons; the power to stop cars for traffic violations gives the police an opportunity to question and search suspicious persons, frequently leading to arrests on nontraffic charges.[42]

Finally, this study of variations in police enforcement policies raises two jurisprudential problems, public control over the exercise of discretion by the police and equal protection of the laws. First, as this study shows almost no evidence to suggest that the police are carrying out *publicly* established enforcement policies, an important set of public policies are being established through a series of "low-visibility" decisions which are either unknown to the public or openly in violation of the "full-enforcement" expectations of the laws and ordinances.[43] (Admittedly, the problem of "low-visibility" is much greater with regard to traffic policy than other areas of law enforcement, but the danger nevertheless remains.) Where these laws cover more than one city (as does a state statute defining traffic violations and the bases for suspension of licenses), variations among police departments also raise the problem of equal protection. Just as a police department which arrests only Negroes is acting unequally, so is a department which chooses to ignore all traffic violations while a neighboring force, applying the same state laws, enforces them vigorously. Our knowledge of the causes of traffic violations and accidents does not yet tell us whether traffic laws should be enforced

[40] Robert P. Shumate, *The Long Range Effect of Enforcement on Driving Speeds: A Research Report* (Washington: International Association of Chiefs of Police, 1960); A. B. Calica, R. F. Crowther, and R. P. Shumate, *Enforcement Effect on Traffic Accident Generation* (Bloomington: University of Indiana Department of Police Administration, 1963).

[41] Daniel Patrick Moynihan, "Public Health and Traffic Safety," *Journal of Criminal Law, Criminology, and Police Science,* LI (May–June, 1960), p. 93; and "The War Against the Automobile," *The Public Interest,* No. 3 (Spring, 1966), pp. 10–26.

[42] See the sources cited in footnote 13, *supra.*

[43] See the sources cited in footnote 15, *supra,* and the works by LaFave and Joseph Goldstein cited in footnote 10, *supra.*

vigorously or leniently to accomplish their purposes. Until all police chiefs in a single legal system adopt a *uniform* policy, however, we are left with a condition of inequality (and thus injustice) by design, with police chiefs making decisions which will determine what proportion of violators will be apprehended and punished. The concept of justice does not tell us *which* violations are most "important," but it does demand that citizens should be treated equally.[44]

[44] See *Ibid.;* Comment, "The Right to Nondiscriminatory Enforcement of State Penal Laws," *Columbia Law Review,* LXI (1961), p. 1103; Note, *Yale Law Journal,* LIX (1950), p. 354.

JAMES Q. WILSON

The Police
and the Delinquent
in Two Cities

The purpose of this chapter is to compare two large American police departments to discover what difference (if any) a high level of "professionalism" makes in the handling of juvenile offenders. If the object of police administration and reform is "professional" standards, it is crucial to know what difference these standards actually make—not simply with respect to honesty but with respect to the quality of justice. The police is one of the agencies that has the most extensive and continuing contact with juvenile delinquents (real or alleged), and no recommendation for the treatment of delinquency should be made without careful consideration of the important differences (if any) in the treatment of delinquents by the police, as well as of the political and organizational situation explaining the differences.

In this chapter, we shall use the terms "juvenile delinquent" and "juvenile offender" to refer to any person under the age of seventeen who commits an act that violates some ordinance or statute. There are, of course, some laws that apply to juveniles only and that, obviously, are

Reprinted from Stanton Wheeler, ed., *Controlling Delinquency* (New York: John Wiley and Sons, 1967), by the permission of the copyright holder, John Wiley and Sons, Inc.

not intended to proscribe "criminal" behavior (laws against loitering, for example, or against truancy from school); we shall, for the most part, confine our comparison to acts which, if committed by an adult, would constitute a crime.

A juvenile is arrested and tried, not for committing a "crime" but for behavior that may eventuate in his being made a ward of the court. Although laws vary from state to state, the common practice (and the practice of the states in which are located the cities that we analyze here) is not to regard proceedings before a juvenile court as criminal but, in the language of one state statute, as intended to "secure for each minor under the jurisdiction of the juvenile court such care and guidance, preferably in his own home, as will serve the spiritual, emotional, mental, and physical welfare of the minor and the best interests of the State; to preserve and strengthen the minor's family ties whenever possible, removing him from the custody of his parents only when his welfare or safety and protection of the public cannot be adequately safeguarded without removal. . . ." In the words of another state statute, delinquent children "shall be treated, not as criminals, but as children in need of aid, encouragement and guidance."

These legal considerations and the customary practices that result from them confer on the authorities considerable discretion in the treatment of juveniles.[1] The police as well as the courts need not and, indeed, do not arrest and punish every child who has committed an act which, if he were an adult, would be a misdemeanor or a felony; the police are generally free to exercise their judgment as to which acts require arrest for the protection of society or the welfare of the child and which acts can be dealt with by other means, including police reprimands, unofficial warnings, or referral to parents or welfare agencies.

The two police departments compared here are those of what we shall call Eastern City and Western City. Both cities have substantially more than 300,000 inhabitants; they are heterogeneous in population and in economic base; both are free of domination by a political machine; and both have a substantial nonwhite population. Western City generally has a mild climate, which probably contributes to rates of crimes against property that are somewhat higher than the rates of Eastern City, where severe winters assist the police in keeping thieves off the streets.

---

[1] On the general problems of police discretion, compare Joseph Goldstein, "Police Discretion Not to Invoke the Criminal Process: Low-Visibility Decisions," *Yale Law Journal*, **69**, 534–594 (March, 1960) and Herman Goldstein, "Police Discretion: The Ideal Versus the Real," *Public Administration Review*, **23**, 140–148 (September, 1963).

THE MEANING OF PROFESSIONALISM

The most important difference between the police of the two cities is that in Western City the police department is highly "professionalized." This does not mean that in Eastern City the police department is wholly corrupt and incompetent; far from it. But as any observer familiar with Eastern City will readily acknowledge, its police officers have been recruited, organized, and led in a way that falls considerably short of the standards set forth in the principal texts. Whether the standards of the texts are right is, of course, another matter. Since the meaning (to say nothing of the value) of professionalism is itself problematical, an effort will be made here to arrive at a general analytical definition and to specify the particular attributes of the police that professionalism implies and how the two police forces differ in these attributes.

A "professional" police department is one governed by values derived from general, impersonal rules which bind all members of the organization and whose relevance is independent of circumstances of time, place, or personality.[2] A nonprofessional department (what will be called a "fraternal" department), on the other hand, relies to a greater extent on particularistic judgments—that is, judgments based on the significance to a particular person of his particular relations to particular others. The professional department looks outward to universal, externally valid, enduring standards; the nonprofessional department looks, so to speak, inward at the informal standards of a special group and distributes rewards and penalties according to how well a member conforms to them.

The specific attributes that are consistent with these definitions include the following ones.

A professional, to a greater extent than a fraternal, department recruits members on the basis of achievement rather than ascriptive criteria. It relies more on standardized formal entrance examinations, open equally to all eligible persons. Thus the professional department recruits not only impartially as to political connections, race or religion; it recruits without regard to local residence. Nonprofessional departments often insist (or laws require them to insist) on recruitment only from among local citizens. Educational standards are typically higher for entrants to professional departments.

Professional departments treat equals equally; that is, laws are enforced

[2] The following definitions are taken from, and treated in greater detail by James Q. Wilson, "The Police and Their Problems: A Theory," *Public Policy,* **12,** 189–216 (1962).

without respect to person. In such departments "fixing" traffic tickets is difficult or impossible and the sons of the powerful cannot expect preferential treatment. Fraternal departments have a less formal sense of justice, either because the system of which they are a part encourages favoritism or because (and this is equally important) officers believe it is proper to take into account personal circumstances in dispensing justice. Concretely, we may expect to find less difference in the professional department between the proportion of white and nonwhite juvenile offenders who are arrested, as opposed to being let off with warnings or reprimands.

Professional departments are less open to graft and corruption and their cities will be more free of "tolerated" illegal enterprises (gambling, prostitution) than will cities with nonprofessional departments.

Professional departments seek, by formal training and indoctrination, to produce a force whose members are individually committed to generally applicable standards. Their training will acquaint them with the writing and teaching of "experts" (that is, of carriers of generalized, professional norms). In fraternal departments, there is less formal training and what there is of it is undertaken by departmental officers who inculcate particularistic values and suggest "how to get along" on the force.

Within the professional department, authority attaches to the role and not to the incumbent to a greater extent than in nonprofessional departments. The essentially bureaucratic distribution of authority within the professional force is necessary because, due to the reliance on achievement, young officers are often promoted rapidly to positions of considerable authority (as sergeants and lieutenants in both line and staff bureaus).[3]

By these tests Western City has a highly professionalized force and Eastern City has not. An observer's first impressions of the two departments suggest the underlying differences: the Western City force has modern, immaculate, and expensive facilities, new buildings, and shiny cars; the officers are smartly dressed in clean, well-pressed uniforms; the routine business of the department is efficiently carried out. In Eastern City, the buildings are old and in poor repair; cars are fewer, many are old and worn; the officers are sometimes unkempt; routine affairs, particularly the keeping of records, are haphazardly conducted by harried or indifferent personnel.

In Western City, three-fourths of the officers were born outside the city and one-half outside the state (this is about the same as the proportion of all males in the city who were born outside the state). In Eastern

---

[3] There is a general tendency for authority to adhere more to the person than to the office in a police force as compared to other kinds of public agencies. See Robert L. Peabody, "Perceptions of Organizational Authority: A Comparative Analysis," *Administrative Science Quarterly*, **6**, 477–480 (March 1962).

City, the vast majority of officers were born and raised within the city they now serve, many in or near neighborhoods in which they now live. In Western City, over one-third of all officers had one year or more of college education; over one-fifth have two years or more; and one-tenth a college degree or better. In Eastern City, the proportion of officers educated beyond high school is far smaller.[4]

In Western City, there was little evidence of gambling or prostitution; Eastern City, while far from "wide open," has not made it difficult for a visitor to find a bookie or a girl. For several years at least, Western City has had a department free from the suspicion of political influence and a court system noted for its "no-fix" policy. In Eastern City, *reports* of influence and fixes are not infrequent (of course, a scholar without the power of subpoena cannot confirm such charges).

The chief of the Western City police department has been a high official of the International Association of Chiefs of Police (IACP); Eastern City's force, by contrast, has been the subject of a special comprehensive report by the IACP, contracted for by the local officials and containing recommendations for extensive reorganization and improvement. In sum, whether judged by subjective impression or objective measure, the police forces of the two cities are significantly different. The crucial question is the consequences of the differences upon the handling of juvenile offenders.

In Western City, justice, on the basis of fragmentary evidence, seems more likely to be blind than in Eastern City. Table 1 shows the percentage of youths of each race arrested or cited in 1962 by Western City's police department for each of the most common offenses. Those not arrested or cited were disposed of, for the most part, by official reprimands. As the table indicates, Negro and white juveniles received remarkably similar treatment for all offenses but two; whites were more frequently arrested than Negroes for aggravated assault, and Negroes more frequently arrested than whites for loitering.

Table 2 gives similar though not precisely comparable information for Eastern City. This table is based on a random sample (1/25) of all juveniles processed by the Eastern City police over the four years

---

[4] These differences are characteristic of entire regions and not simply of the two departments here studied. In one study it was found that almost 90 per cent of the officers in police departments in the Pacific states, but only about two-thirds of those in New England and North Atlantic states, had a high school education. Similarly, 55 per cent of those from the Pacific states, but only about 18 per cent of those from New England and North Atlantic states, had attended college. See George W. O'Connor and Nelson A. Watson, *Juvenile Delinquency and Youth Crime: The Police Role* (Washington, D.C.: International Association of Chiefs of Police, 1964), pages 78–79.

TABLE 1   PROPORTION OF SUSPECTED JUVENILE OFFENDERS ARRESTED OR
CITED, BY RACE, FOR SELECTED OFFENSES IN WESTERN
CITY (1962)

Offense	Total Offenses		Per Cent Arrested or Cited	
	White	Negro	White	Negro
Robbery	19	105	100.0	92.4
Aggravated assault	9	61	78.8	55.4
Burglary	199	331	87.9	92.8
Auto theft	124	142	93.6	86.6
Larceny	459	1119	56.2	56.6
Loitering	504	829	12.5	20.0
Drunk and disorderly	151	343	39.1	34.7
Malicious mischief	213	216	33.8	37.5
Assault and battery	93	306	58.1	65.3
Total	1771	3452	46.5	50.9

since the juvenile bureau began keeping records. The data on offense
and disposition were taken from cards for individual juveniles; thus, the
figures in Table 2 show what proportion of *juveniles,* by race, were taken
to court for various offenses, while the figures in Table 1 show what
proportion of *juvenile offenses* (many juveniles being counted more than
once), by race, resulted in a court disposition. Despite the lack of strict

TABLE 2   PROPORTION OF JUVENILES (IN 1/25 SAMPLE OF ALL THOSE
PROCESSED) TAKEN TO COURT, BY RACE, FOR SELECTED
OFFENSES IN EASTERN CITY (1959–1961)

Offense	Total Offenses		Per Cent Taken to Court	
	White	Negro	White	Negro
Assaults	26	12	11.5	25.0
Burglary	34	4	11.8	100.0
Auto theft	7	3	42.8	66.7
Larceny	98	27	24.5	52.0
Drunk and disorderly	33	4	0.0	0.0
Malicious mischief	69	9	4.4	0.0
Incorrigible	20	4	40.0	100.0
Total	287	63	15.7	42.9

comparability, the differences are worth consideration. Although, in Western City, there was little difference in the probability of arrest for whites as compared to Negroes, in Eastern City the probability of court action (rather than warnings or reprimands) is almost three times higher for Negroes than for whites.

HANDLING THE DELINQUENT

The two police departments are systematically different both in their treatment of delinquents and in the way the members think and talk about delinquents; paradoxically, the differences in behavior do not correspond to the verbal differences. Interviews with approximately half the officers (selected at random) assigned to the juvenile bureaus of the police departments of Eastern and Western Cities reveal that Western City's officers have more complex attitudes toward delinquency and juveniles than their colleagues of Eastern City. The former's attitudes, at least superficially, tend to be less moralistic, less certain as to causal factors, more therapeutic, and more frequently couched in generalizations than in anecdotes. Eastern City's officers, by contrast, are more likely to interpret a problem as one of personal or familial morality rather than social pathology, to urge restrictive and punitive rather than therapeutic measures, to rely on single explanations expressed with great conviction and certainty, and to confine discussions of juveniles almost exclusively to anecdotes and references to recent episodes than to generalizations, trends, or patterns.[5]

The behavior of the officers with respect to juveniles tends to be the opposite of what we might expect from their expressed sentiments. In Western City, the discretionary powers of the police are much more likely than in Eastern City to be used to restrict the freedom of the juvenile: Western City's officers process a larger proportion of the city's juvenile population as suspected offenders and, of those they process, arrest a larger proportion.

Table 3 shows the total number of juveniles processed in 1962 by the departments of Eastern and Western City, the rate per 100,000 juveniles in the populations of the two cities, and the percentage (and rate) arrested or cited. By "processed" is meant that the youth came in contact with

[5] Compare these dichotomous attitudes with those classified in Walter B. Miller, "Inter-Institutional Conflict as a Major Impediment to Delinquency Prevention," *Human Organization*, **17** (3), 20–23, and Harold L. Wilensky and Charles N. Lebeaux, *Industrial Society and Social Welfare* (New York: Russell Sage Foundation, 1958), pp. 219–228.

TABLE 3 NUMBER AND RATE OF JUVENILES PROCESSED AND ARRESTED IN WESTERN CITY AND EASTERN CITY (1962)

	Western City	Eastern City
Total juveniles processed (all offenses)	8,331	6,384
Rate per 100,000 children[a]	13,600	6,380
Number of juveniles arrested or cited		
(all offenses)	3,869	1,911
Per Cent arrested or cited	46.8	30.0
Rate per 100,000 children	6,365	1,910
Total juveniles processed, less those		
charged with loitering	6,685	6,384
Rate per 100,000 children	10,900	6,380
Number of juveniles arrested or cited,		
less loiterers	3,446	1,911
Per Cent arrested or cited	51.6	30.0
Rate per 100,000 children	5,623	1,910

[a] Rate is based on number of children, ages six through sixteen, in population of city according to the 1960 census of population.

the police in a manner that required the latter to take official cognizance; a report was filed or a record entry made on the ground that the police had reasonable cause to believe that the youth had engaged in, or was a material witness to, acts which brought him under provisions of the state statutes. By "arrested or cited" is meant that the police brought formal action against the juvenile, either by taking him into custody and thence turning him over officially to the courts or to the probation officers or referees who can make a preliminary disposition, or by issuing an order or citation requiring him to appear before a court or official of the probation department. Such dispositions should be contrasted with all others in which the possibility of punitive action does not exist: officially reprimanding and releasing the child, referring him to another agency, returning him to his parents, and so forth. In short, the proportion arrested or cited is the proportion of all juveniles, suspected of having committed any offense, for whom the police make official punitive action a possibility—although not a certainty.

The rate of juveniles (Table 3) processed for *all* offenses by Western City's police was more than twice as great as the rate in Eastern City (13,600 per 100,000 as opposed to 6,380 per 100,000) and, of those processed, the proportion arrested or cited was more than 50 per cent greater in Western than in Eastern City (46.8 per cent as opposed to 30 per cent). However, the laws of the two cities (and consequently the number of grounds on which juveniles can be processed by the police) differ, and therefore these raw figures must be modified by eliminating all juveniles processed for offenses unique to one place. The only type of offense that involved more than 1 per cent of the juveniles was an antiloitering ordinance in effect in Western City but not in Eastern City. (The fact that Western City *has* such an ordinance—that forbids persons under the age of eighteen from loitering unaccompanied by a parent in public places between 10:00 P.M. and sunrise—is, it can be argued, in itself a manifestation of the difference in the conception of justice prevailing in each city. Not only does Eastern City not have such an ordinance but the head of the juvenile bureau at the time of this research was opposed to its adoption.) Table 3 gives the adjusted figures after deleting from Western City's totals all juveniles processed or arrested for violation of the antiloitering ordinance; yet both the processing rate and the arrest rate remain over 50 per cent higher than in Eastern City.

In short, the young man or woman in Western City is one and one-half to two times as likely to come into contact with the police and, once in contact with the police, one and one-half times as likely to be arrested or cited rather than reprimanded or referred. One explanation of the contrast might be that, because of circumstances over which the police have no control, all the people there—the old as well as the young—are more likely to commit criminal acts. The more favorable climate, for example, might well explain why there were more crimes against property in Western City than in Eastern City. It can be argued that young people who are not professional thieves are even more likely than adults to be deterred by wind, snow, and freezing temperatures from stealing cars or breaking into hardware stores. Furthermore, Western City has a higher proportion of Negroes than Eastern City. If, in fact, Western City's youths are "more criminal" or have more opportunity for criminal acts, then differences in processing and arrest rates might reveal nothing about police attitudes or community norms.

In an effort to evaluate this objection, a more detailed comparison of crime and arrest rates for both juveniles and adults is given in Table 4. For each of the "Part I" offenses, the seven most serious offenses as defined by the FBI, overall crime rates (that is, offenses known to the police and arrest rates for both adults and juveniles) are given for both cities for 1962. In sum, the crime rates of the two cities are remarkably similar, although some considerable disparities are concealed in the totals. As one might predict, Western City has a substantially higher crime

TABLE 4 CRIME RATES AND ADULT AND JUVENILE ARREST RATES
PER 100,000 POPULATION FOR WESTERN CITY AND
EASTERN CITY, 1962, BY MAJOR OFFENSE

Offense	Crime Rate		Adult Arrest Rate		Juvenile Arrest Rate		Ratio[a]	
	Western	Eastern	Western	Eastern	Western	Eastern	Adult	Juvenile
Homicide	8	7	10	11	3	0.2	0.9	17.0
Forcible rape	17	15	14	14	36	13	1.0	2.8
Robbery	167	104	115	71	197	63	1.6	3.1
Aggravated assault	105	117	85	107	69	73	0.8	0.9
Burglary	957	566	188	118	860	334	1.6	2.6
Larceny	458	420	576	240	1,580	664	2.4	2.4
Auto theft	357	855	71	127	450	316	0.6	1.4
Total	2,069	2,084	1,059	688	3,195	1,463	1.5	2.2

[a] These are the ratios of arrest rates in Western City to arrest rates in Eastern City for adults and juveniles. The ratio was calculated by dividing the Western City rate by the Eastern City rate. Values in excess of 1.0 are measures of the degree to which Western City rates exceed Eastern City rates.

rate for robbery and burglary; unexpectedly, Eastern City has a substantially higher rate for auto theft. Crimes against the person—homicide, forcible rape, and aggravated assault—are quite similar in the two cities, crimes of passion being unfortunately less inhibited by adverse weather, probably because so many of them occur indoors. The arrest rates are a different story: the rates for both adults and juveniles are higher in Western City than in Eastern City, but the difference is greatest for the juveniles.

Western City's police arrest a greater proportion of the population than do Eastern City's but whereas the former's rate is 50 per cent higher for adults, it is over 100 per cent higher for juveniles. The last two columns in Table 4 summarize these differences by showing, for both adults and juveniles, the ratio between Western and Eastern City rates of arrest for each offense. Only for aggravated assault were the rates for juveniles of Eastern City higher than those of Western City; for all other offenses, Western City's rates were generally from 1.4 to 3.1 times greater. Particularly striking is the fact that, although the *auto theft rate* was over twice as high in Eastern City, the *juvenile arrest rate* for auto theft was 40 per cent greater in Western City.

Thus, a juvenile in Western City is far less likely than one in Eastern City to be let off by the police with a reprimand. What the data indicate, interviews confirm. Police officers, social workers, and students of delinquency in Eastern City agree that the police there are well-known for what is called by many the "pass system." Unless the youth commits what the police consider a "vicious" crime—brutally assaulting an elderly person, for example, or engaging in wanton violence—he is almost certain to be released with a reprimand or warning on his first contact with the police and quite likely to be released on the second, third, and sometimes even on the fourth contact. It must be said that the juvenile officer who handles the case may consult a card file in his station showing previous police contacts for all juveniles in the precinct; a "pass" is not given out of ignorance.

The account of one Eastern City juvenile officer is typical of most accounts:

Most of the kids around here get two or three chances. Let me give you an example. There was this fellow around here who is not vicious, not, I think, what you'd call bad; he's really sort of a good kid. He just can't move without getting into trouble. I don't know what there is about him . . . I'll read you his record. 1958—he's picked up for shoplifting, given a warning. 1958— again a few months later was picked up for illegal possession [of dangerous weapons]. He had some dynamite caps and railroad flares. Gave him another warning. 1959—the next year he stole a bike. Got a warning. 1960—he broke into some freight cars. [Taken to court and] continued without a finding [that is, no court action] on the understanding that he would pay restitution. Later the same year he was a runaway from home. No complaint was brought against him. Then in 1960 he started getting into some serious stuff. There was larceny and committing an unnatural act with a retarded boy. So, he went up on that one and was committed to [the reformatory] for nine months. Got out. In 1962 he was shot while attempting a larceny in a junk yard at night . . . Went to court, continued without a finding [that is, no punishment]. Now that's typical of a kid who just sort of can't stay out of trouble. He only went up once out of, let me see . . . eight offenses. . . . I wouldn't call him a bad kid despite the record . . . the bad kids: we don't have a lot of those.

In the Eastern City, there are, of course, officers who have the reputation for being "tough." The "toughness" may be manifested, however, not so much in more frequent court appearances of youths, but in the greater ease of getting information. "Tough" and "soft" officers work as teams, the latter persuading juveniles to talk in order to save them from the former. In any case, the net effect of police discretion in Eastern City is unambiguous; only 17.5 per cent of the first offenders included in a 1/25 random sample of all juveniles processed over a four year period by the police department were referred to court. Indeed, Eastern City's officers occasionally mentioned that it was their understanding that officers "in the West" made arrests more frequently than they; Western City's officers sometimes observed that they had been told that officers "in the East" made arrests less frequently than they.

Observation of the operation of the two departments provided considerable evidence of the effect of the preceding on the day-to-day practice of police work. While cruising the city in patrol cars, Western City's officers would frequently stop to investigate youths "hanging" on street corners; the officers would check the youths' identification, question them closely, and often ask over the radio if they were persons for any reason wanted at headquarters. In Eastern City, officers would generally ignore young persons hanging around corners except to stop the car, lean out, and gruffly order them to "move along." "Sweeping" or "brooming" the corners was done with no real hope on the part of the police that it would accomplish much ("they'll just go around the block and come right back here in ten minutes") but they would ask, "what else can you do?"

Technically, of course, an officer in either city who takes a person into custody on the street is required by law to bring him to police headquarters or to a station house and to initiate a formal procedure whereby an arrest is effected, charges stated, certain rights guaranteed, and, if necessary, physical detention effected. In fact, and particularly with respect to juveniles, police officers sometimes take persons directly to their homes. In Eastern City this procedure is, in my judgment (naturally no conclusive evidence is available), much more common than in Western City.

THE CORRELATES OF DISCRETION

If, at least in this one case, a "professionalized" police department tends to expose a higher proportion of juveniles to the possibility of court action, despite the more "therapeutic" and sophisticated verbal formulas of its officers, it is important to ask why this occurs. Many reasons suggest themselves but, since this research is limited to an intensive examination of two departments, with a more cursory examination of two others, it is impossible to say how much of the variation in arrest rates can be accounted for by any single circumstance or by all circumstances together, or whether in other cities different relationships might be found. However, a rather strong argument can be made that, at the very least, the relationship is not accidental and, further, that professionalism itself in a variety of ways contributes to the result. Finally, what at first seems a paradox— the discrepancy between ideology and behavior—is not in fact a paradox at all, but simply the differing expression of a single state of affairs.

Certain structural and procedural dissimilarities undoubtedly account for some of the differences in arrests. In Eastern City the juvenile officer on the police force is also the prosecuting officer: he personally prepares and presents the case in court against the juvenile. In Western City, the juvenile officer (who, as in Eastern City, takes charge of the juvenile after a patrolman or detective has "brought him in") prepares an initial report but sends the report and, if detention seems warranted, the child himself, to an independent probation department which determines whether the suspect should be taken before the judge. In effect, Western City officers can "pass the buck," and even if the case goes to court, the officer himself only rarely is required to appear in court. In Eastern City, the police are involved right up to the moment when the judge actually makes a disposition, a police appearance being always required if there is to be a court hearing. Moreover, the probation department is not independent, but is an arm of the court which acts only *after* a court appearance. As a result of these arrangements, Eastern City's officers may have an incentive not to send the child to court because it requires more work; to Western City's officers, on the other hand, initiating a court appearance is relatively costless.

But such considerations do not explain why the *arresting* officer (who in most cases is *not* the juvenile officer who makes the ultimate disposition or the court appearance) should be less likely to make an arrest in one city than the other—unless, of course, there is some social pressure from juvenile officers in Eastern City to keep down the arrest rate. There is such pressure but, as will be shown, it does not come from juvenile officers but from the force as a whole.

It may be, of course, that the force as a whole is influenced by its perception of the probability that the court will actually punish the suspect, although it is by no means clear which way this influence might work. On the one hand, a lenient court might prove discouraging to the police, leading them to conclude that the "kid will get off anyway, so why should I go to the trouble of making an arrest?" On the other hand, a lenient court could as easily lead officers to argue that, since the kid will be "let off," there is no real danger to the suspect and, therefore, he may as well be arrested, whatever the merits of the case, as a way of throwing a harmless but perhaps useful "scare" into him.

This need not be solved, however, because the officers in both cities perceive the court, together with the probation authorities, as "excessively" lenient. And with good reason: in Eastern City, even though the police take only about 17.5 per cent of all first offenders to court, only a third of these get any punishment at all and less than a tenth are sent to a reformatory. Those not committed to a reformatory are given suspended sentences or placed on probation. In sum, *only 1.6 per cent* of first offenders see the inside of a correctional institution. In Western City, comparable figures are difficult to assemble. Generally speaking, however, the police are correct in their belief that only a small fraction of the youths they refer to the probation department will be sent on to court and that, of these, an even smaller fraction will be committed to a correctional institution. Of the more than eight thousand juveniles processed by the police in 1962, slightly less than half were referred, by arrest or citation, to the probation department. Of the juveniles referred, about one-third were ordered by the probation department to make a court appearance; of these, about one-sixth were sent to a public institution. In sum, *only about 2.8 per cent* of the juveniles processed by the police were in some way confined. The differences in the probability of punishment in the two cities were so small as to make them a negligible influence on police behavior.

Far more important, it seems to me, than any mechanical differences between the two departments are the organizational arrangements, community attachments, and institutionalized norms which govern the daily life of the police officer himself, all of which might be referred to collectively as the "ethos" of the police force. It is this ethos which, in my judgment, decisively influences the police in the two places. In Western City, this is the ethos of a *professional* force; in Eastern City, the ethos of a *fraternal* force.

Western City's police officer works in an organizational setting which is highly centralized. Elaborate records are kept of all aspects of police work: each officer must, on a log, account for every minute of his time on duty; all contacts with citizens must be recorded in one form or another; and automatic data-processing equipment frequently issues detailed reports on police and criminal activity. The department operates out of a single headquarters; all juvenile offenders are processed in the office of the headquarters' juvenile bureau in the presence of a sergeant, a lieutenant, and, during the day shift, a captain. Dossiers on previously processed juveniles are kept and consulted at headquarters. Arresting officers bring all juveniles to headquarters for processing and their disposition is determined by officers of the juvenile bureau at that time.

In Eastern City, the force is highly decentralized. Officers are assigned to and, sometimes for their whole career, work in precinct station houses. Juvenile suspects are brought to the local station house and turned over to the officer of the juvenile bureau assigned there. These assignments are relatively constant: a patrolman who becomes a juvenile officer remains in the same station house. The juvenile officer is not supervised closely or, in many cases, not supervised at all; he works in his own office and makes his own dispositions. Whatever records the juvenile officer chooses to keep—and most keep some sort of record—is largely up to him. Once a week he is required to notify the headquarters of the juvenile bureau of his activities and to provide the bureau with the names and offenses of any juveniles he has processed. Otherwise, he is on his own.[6]

The centralized versus the decentralized mode of operations is in part dictated by differences in size of city—Eastern City has a larger population than Western City—but also in great part by a deliberate organizational strategy. Western City at one time had precincts, but they were abolished by a new, "reform" police chief as a way of centralizing control over the department in his hands. There had been some scandals before his appointment involving allegations of police brutality and corruption which he was determined would not occur again. Abolishing the precincts, centralizing the force, increasing the number and specificity of the rules and reporting procedures, and tightening supervision were all measures to achieve this objective. These actions all had consequences, many of them perhaps unintended, upon the behavior of the department. Officers felt the pressure: they were being watched, checked, supervised, and reported on. The force was becoming to a considerable extent "bureaucratized"— behavior more and more was to involve the nondiscretionary application of general rules to particular cases.[7] Some officers felt that their "productivity" was being measured— number of arrests made, citations written, field contact reports filed, and suspicious persons checked. Under these circumstances, it would be surprising if they did not feel they ought to act in such a way as to minimize any risk to themselves that might arise, not simply from being brutal or taking graft, but from failing to "make pinches" and "keep down the crime rate." In short, organizational measures intended to insure that police behave properly with respect to nondiscretionary matters (such as taking bribes) may also have the effect (perhaps unintended) of making them behave differently with respect to matters over which they *do* have discretion. More precisely, these measures tend to induce officers to convert discretionary to nondiscretionary matters—for example, to treat juveniles according to rule and without regard to person.

[6] The juvenile bureau of the Eastern City police department was only created after community concern over what appeared to be a serious incident involving a juvenile "gang" compelled it. The police commissioner at the time was reported to oppose the existence of such a bureau on the revealing grounds that "each beat officer should be his own juvenile officer." The fraternal force apparently resisted even the nominal degree of specialization and centralization represented by the creation of this bureau. This, again, is also a regional phenomenon. Over 80 per cent of the police departments in Pacific states, but less than 58 per cent of those in New England states, have specialized juvenile units. O'Connor and Watson, *op. cit.*, p. 84.

[7] Compare the causes and consequences of bureaucratization in an industrial setting in Alvin W. Gouldner, *Patterns of Industrial Bureaucracy* (Glencoe, Ill.: Free Press, 1954).

In Eastern City the nonprofessional, fraternal ethos of the force leads officers to treat juveniles primarily on the basis of personal judgment and only secondarily by applying formal rules. Although the department has had its full share of charges of corruption and brutality, at the time of this research there had been relatively few fundamental reforms. The local precinct captain is a man of great power; however, he rarely chooses to closely supervise the handling of juvenile offenders. His rules, though binding, are few in number and rarely systematic or extensive.

In Western City, the juvenile officers work as a unit; they meet together every morning for a line-up, briefing, and short training session; they work out of a common headquarters; they have their own patrol cars; and they work together in pairs. In Eastern City most, though not all, precincts have a single juvenile officer. He works in the station house in association with patrolmen and detectives; he has no car of his own, but must ride with other officers or borrow one of their cars; he rarely meets with other juvenile officers and there is practically no training for his job or systematic briefing while on it. In Western City, the juvenile officer's ties of association on and off the job are such that his fellow juvenile officers are his audience. He is judged by, and judges himself by, their standards and their opinions. In Eastern City, the relevant audience is much more likely to be patrolmen and detectives. In Western City, the primary relations of the juvenile officer are with "professional" colleagues; in Eastern City, the relations are with fraternal associates.

Eastern City's juvenile officer feels, and expresses to an interviewer, the conflicting and ambivalent standards arising out of his association with officers who do not handle juveniles. On the one hand, almost every juvenile officer in Eastern City complained that patrolmen and detectives did not "understand" his work, that they regarded him as a man who "chased kids," that they "kissed off" juvenile cases onto him and did not take them seriously, and that they did not think arresting a "kid" constituted a "good pinch." These attitudes might, in part, be explained by the patrolmen's reluctance to bring a juvenile into the station, even if they could then turn him over to the juvenile officer on duty; bringing the boy in meant bringing him in in front of their fellow patrolmen in the squad room of the station house. One patrolman's views on this were typical:

A delinquent is not a good pinch—at least not for most officers. You get ribbed a lot and sort of ridiculed when you bring a kid in. Sort of grinds you down when you bring a kid in and the other officers start telling you, "Hey, look at the big man, look at the big guy with the little kid, hey, can you handle that kid all by yourself?" You get a little ribbing like that and finally you don't bring so many kids in for pinches.

Instead, the patrolmen or detectives often simply refer the juvenile's name to the juvenile officer and let the officer go out and handle the case from investigation to arrest. This not only places a larger work load on the juvenile officer; it places it on him under conditions that do not reward effective performance. He is given the "kid stuff" because patrolmen do not feel rewarded for handling it; at the same time, the patrolman lets it be known that he does not feel the juvenile officer ought to get much credit, either. At the same time, almost all patrolmen interviewed felt that the authorities, including in most cases the juvenile officer himself, were "too easy" on the kids. But this generalized commitment to greater punitiveness, although widely shared in Eastern City, rarely—for reasons to be discussed later—determines the fate of any particular juvenile. This being the case, the juvenile officer in Eastern City seems to allow his behavior to be influenced by associates, insofar as it is influenced by them at all, in the direction of permissive treatment.

Western City's juvenile officers, by contrast, are more insulated from or less dependent on the opinion of patrolmen and detectives. And the latter, when taking a juvenile into custody, can bring him to a central juvenile bureau staffed only by juvenile officers, rather than to a precinct station filled with fellow patrolmen and detectives. Neither juvenile officers nor arresting patrolmen are, in Western City, as directly exposed to or dependent upon the opinions of associates concerning whether a juvenile arrest is justified.

Even if Western City's officers should be so exposed, however, it is likely that they would still be more punitive than their counterparts in Eastern City. In Western City, the officer, both in and out of the juvenile bureau, is recruited and organized in a way that provides little possibility of developing a strong identification with either delinquents in general or with delinquents in some particular neighborhood. He is likely to have been raised outside the city and even outside the state; in many cases he was recruited by the representatives of the force who canvass the schools of police administration attached to western and midwestern universities. In only *one* case in Western City did I interview a juvenile officer who, when asked about his own youth, spoke of growing up in a "tough" neighborhood where juvenile gangs, juvenile misbehavior, and brushes with the police were common. There were, on the other hand, only one or two of Eastern City's officers who had *not* come from such backgrounds: they were almost all products not only of local neighborhoods but of neighborhoods where scrapes with the law were a common occurrence.

The *majority* of Eastern City's officers were not only "locals," but locals from lower or lower-middle-class backgrounds. Several times officers spoke of themselves and their friends in terms that suggested that the transition between being a street-corner rowdy and a police officer was not very abrupt. The old street-corner friends that they used to "hang" with followed different paths as adults but, to the officers, the paths were less a matter of choice than of accident, fates which were legally but not otherwise distinct. The officers spoke proudly of the fights they used to have, of youthful wars between the Irish and the Italians, and of the old gangs, half of whose alumni went to the state prison and the other half to the police and fire departments. Each section of the city has great meaning to these officers; they are nostalgic about some where the old life continues, bitter about others where new elements—particularly Negroes—have "taken over."

The *majority* of Western City's officers who were interviewed, almost without exception, described their own youth as free of violence, troubles with the police, broken homes, or gang behavior. The city in which they now serve has a particular meaning for only a very few. Many live outside it in the suburbs and know the city's neighborhoods almost solely from their police work. Since there are no precinct stations but only radio car routes, and since these are frequently changed, there is little opportunity to build up an intimate familiarity, much less an identification, with any neighborhood. The Western City police are, in a real sense, an army of occupation organized along paramilitary lines.

It would be a mistake to exaggerate these differences or to be carried away by neighborhood romanticism that attaches an undeservedly high significance to the folklore about the "neighborhood cop" walking his rounds—king of the beat, firm arbiter of petty grievances, and gruff but kind confidant of his subjects. The "oldtime beat cop," as almost all the Eastern City's officers are quick and sad to admit, is gone forever. But even short of romanticism, the differences remain and are important. Except for the downtown business district and the skid row area, there are no foot patrolmen in Western City; in Eastern City, in all the residential areas with high crime rates, officers walk their beats. Furthermore, the station houses in Eastern City receive a constant stream of local residents who bring their grievances and demands to the police; in Western City the imposing new police headquarters building is downtown and has no "front desk" where business obviously can be transacted. Although visitors are encouraged, upon entering the ground floor one is confronted by a bank of automatic elevators. Finally, officers on duty in Eastern City eat in diners and cafés in or close to their routes; in Western City officers often drive several miles to a restaurant noted for its reasonably-priced food rather than for its identification with the neighborhood.

These differences in style between the two police departments can perhaps be summarized by saying that in Western City the officer has a generalized knowledge of juveniles and of delinquency and that, although he, of course, becomes familiar with some children and areas, that generalized knowledge—whether learned in college, from departmental doctrine, from the statute books, or from the popular literature on juvenile behavior—provides the premises of his decisions. He begins with general knowledge and he is subjected to fewer particularizing influences than his counterpart in Eastern City. In Eastern City, the officer's knowledge or what he takes to be his knowledge about delinquency, crime, and neighborhood affairs is, from the first, specific, particular, indeed, *personal,* and the department is organized and run in a way that maintains a particularist orientation toward relations between officer and officer and between police and citizens.[8]

This Eastern City ethos exists side by side with the general moral absolutism of police attitudes toward delinquency *in general.* When asked about the cause, extent, or significance of delinquency *generally,* the officers usually respond, as has been indicated, with broad, flat, moral indictments of the modern American family, overly-indulged youth, weakened social bonds, corrupting mass media of communication, and pervasive irreligion and socialism. When the same officers are asked about delinquency in *their precinct,* they speak anecdotally of particular juveniles engaging in particular acts in particular circumstances, in dealing with whom they apply, not their expressed general moral absolutes, but their particular knowledge of the case in question and some rough standard of personal substantive justice.

The one striking exception arises when Negroes are involved. The white officer is not in any kind of systematic communication with Negroes; the Negro is the "invader," and—what may be statistically true—more likely to commit crimes. The officer sees the Negro as being often more vicious, certainly more secretive, and always alien. To the policeman of Eastern City, the Negro has no historical counterpart in his personal experience and, as a result, the Negro juvenile is more likely than the white to be treated, in accord not with particularist standards, but with the generalized and absolutist attitudes which express the officer's concern for the problem "as a whole."

[8] The findings of O'Connor and Watson are consistent with this argument. They discovered that officers in Pacific police departments tended to have "tougher" attitudes toward the *means* to be employed in handling juvenile offenders than officers in New England or North Atlantic departments. The former were more likely to favor transporting juveniles in marked rather than unmarked police cars, to favor having a curfew, to oppose having the police get involved in community affairs concerning youth matters, to oppose destroying the police records of juveniles after they become adults, and to oppose having the police try to find jobs for juveniles who come to their attention. (*Op. cit.,* pp. 91–97, 115–127.)

One reason for the apparently higher proportion of arrests of Negro compared to white juveniles in Eastern City may have nothing to do with "prejudice." In addition to being perceived as an "alien," the Negro offender is also perceived as one who "has no home life." Eastern City officers frequently refer to (and deplore) the apparent weakness of the lower-class Negro family structure, the high proportion of female heads of households, and the alleged high incidence of welfare cases (notably Aid to Dependent Children). If a fraternal force is concerned as much with the maintenance of family authority as with breaking the law and if referring the child to the home is preferred to referring him to court, then the absence (or perceived absence) of family life among Negroes would lead to a greater resort to the courts.

Western City's officer, acting on essentially general principles, treats juveniles with more severity (concern with distinctions of person is less, though by no means entirely absent) but with less discrimination. Negroes and whites are generally treated alike and both are treated more severely. Because the officer in this city is more likely to be essentially of middle-class background and outlook and sometimes college educated as well, he is much more likely to be courteous, impersonal, and "correct" than the Eastern City officer.

These differences in organizational character are reinforced by the political and civic institutions of the cities. Eastern City has been governed for decades by "old-style" politics; personal loyalties, neighborhood interests, and party preferment are paramount. Western City is preeminently a "good government" community in all respects. The nonpartisan city council and administration have, by and large, made every effort to make the management of the community honest and efficient. In this, they have been supported by the press and by business and civic groups. Western City's police chief enjoys strong support in his determined effort to maintain and extend professional standards in the force. Eastern City's chief is, of course, expected to perform creditably and avoid scandal but the whole tenor of the city's political life provides little evidence that anything more than routine competence is required or even wanted.

The differences in the ethos of both the police and the political institutions of the two communities may have a common prior cause. Even though Western City has a much larger proportion of Negroes than Eastern City, the city as a whole is slightly more "middle class" (one-fifth of the families have incomes over $10,000 a year compared to less than one-seventh in Eastern City) and significantly less "European" (less than one-fifth of the population is of foreign stock compared to nearly one-half in Eastern City). In Western City, about half the dwelling units are owner-occupied; in Eastern City, only one-third are. Three times as many households have second cars in Western City as in Eastern City.

It is not unreasonable to assume that community expectations influence the behavior of both politicians and police officers. A somewhat more middle-class community will expect more vigorous law enforcement just as—and for the same reason—it expects honest, efficient municipal management. The universalistic norms of the professional force and the particularistic norms of the fraternal force are consistent with and supported by community values.[9] (This is easily exaggerated. The differences in class composition of the two cities are not great; further, before the 1950's Western City had a lax police force and city government.)

SOME POLICY IMPLICATIONS

If this analysis is substantially correct, it will suggest several things to the policy maker. First, decisions with respect to "professionalizing" big-city police forces should not be taken without consideration of their effect on the justice meted out in discretionary cases. It is not possible—as we should have known all along—simply to make a police force "better"; these questions must first be answered: "better for what?" and "better for whom?" Students of public administration have argued long and correctly that "efficiency" or "management" can rarely be "improved" without some effect on the substantive goals of the organization. It should not be surprising that, in police departments, as elsewhere, means are almost never purely instrumental but have their consequences upon ends.

The second implication directly concerns the problem of juvenile delinquency. The training of a police force apparently alters the manner in which juveniles are handled. A principal effect of the inculcation of professional norms is to make the police less discriminatory but more severe. As a political scientist, I cannot pretend to know whether, from the standpoint of "solving" or "treating" the "delinquency problem," this is good or bad. I find it difficult, however, to believe that the issue is settled. Plausible arguments can be advanced, I am confident, to the effect that certain, swift punishment—in this case, certain, swift referral to a court agency—is an excellent deterrent to juvenile crime. Youths are impressed early, so the argument might go, with the seriousness of their offenses and the consequences of their actions. Equally plausible arguments can no doubt be adduced that arresting juveniles—particularly first offenders—tends to confirm them in deviant behavior: it gives them the status,

[9] Differences in political ethos may parallel differences in police ethos. Compare the differences between professional, generalized norms and fraternal, particularist norms with the differences between the Anglo-Saxon middle class and the "immigrant" political ethos as described in Edward C. Banfield and James Q. Wilson, *City Politics* (Cambridge: Harvard University Press, 1963), especially Chapters 3, 11, 16, and 22.

in the eyes of their gang, of "tough guys" who have "been downtown" with the police; it throws them into intimate contact with confirmed offenders, where presumably they become "con-wise" and learned in the tricks of the thievery trade; and, somewhat contradictorily since sentencing is rarely severe, it gives them a contempt for the sanctions open to society.

There is probably some truth in both arguments. Different strategies may work with different juveniles, and these differences may cluster along lines of class, ethnicity, or family background.

Whether Western City or Eastern City has been more effective in reducing or preventing juvenile crime is hard to say. No agency can compile statistics on crimes that were *not* committed and the police cannot, of course, specify which known but unsolved crimes were committed by juveniles. Trends in the crime rate may have more to do with population changes than with police activity. The only available comparative indicator is the clearance rate—the proportion of offenses known to the police "cleared" by arrest. These rates, however, are typically calculated differently in different cities and are based on highly subjective judgments as to how many and what kinds of offenses are "cleared" by an arrest. It would be misleading to use these rates to measure "police effectiveness."

The policy maker, searching for a way to adapt police practice to the problem of delinquency, could properly ask whether certain aspects of the professional force contribute more than others to the results described here. To that, no clear answer can be given. There is some evidence that among smaller cities, upper-class communities have police forces which *process* more juveniles but *arrest* fewer than police forces in nearby lower- or lower-middle-class communities.[10] Thus, the size of the city may be an important variable: perhaps in the *small* community with an upper-class, "good government" ethos, more offenses are called to the attention of the police—hence the larger number of juveniles processed—while the smaller, more intimate nature of the community encourages the police to rely more on turning offenders over to their parents. Since children of upper-class families are less likely than those of lower-class families to commit major crimes, such as burglary and larceny, their referral by the police to their parents involves less sacrifice of professional norms. On the other hand, in *large* communities with professional, "good government" standards, the greater impersonality of the police may mean that, although citizens report offenses, they do not expect—and do not get—police referral of children to the parents rather than the courts.

Another variable that may effect differences in police behavior is age.

[10] See Nathan Goldman, *The Differential Selection of Juvenile Offenders for Court Appearance* (New York: National Research and Information Center of the National Council on Crime and Delinquency, 1963), especially pp. 48–124.

Western City's officers are, on the whole, younger than Eastern City's; further, among the latter, one can observe a tendency for the younger officers to act more like their counterparts in Western City. There is no direct evidence on this point, but further research might show that younger officers are more zealous than the older (that is, more likely to act on the basis of organizational rule than personal judgment), that they take their jobs more seriously, that they are likely to be better educated, and that all these qualities—zeal, seriousness, and education—combine to make the officer more likely to investigate suspicious circumstances, process a large number of alleged delinquents, and, of those processed, arrest a larger proportion.

The "age" of the force as a whole may further contribute to the differences. The Western City force is not only young in the average age of its members, but "young" in the sense that it was within the last ten years "reformed" and reorganized and that zealous, recently promoted officers occupy positions of authority in the department. Another big-city police department that I visited (let me call it Center City) that also has high professional standards and a record free of any major scandal involving graft or collusion did not (in comparison to Western City) process as large a proportion of the city's juveniles or arrest as large a proportion of those processed. (In Center City data are not collected in such a way as to make them exactly comparable with Western City data.) The strongest impression an observer carries away from a prolonged visit to the Center City department is that the force, while honest and competent, has lost its sense of zeal. It was "reformed" over twenty years ago after a series of major scandals. The young officers who then rose rapidly to positions of influence and who presumably for a time gave to the force a new vigor, have grown older. The tightness of supervision so characteristic of the Western City force is absent in Center City: perhaps over the years it has simply grown slack. The city remains "closed" to vice and gambling but, with respect to juveniles, there is a greater propensity to "reprimand and release" than to arrest or cite.

For these two reasons—size of city and "age" of force—and perhaps for many others as well, the conclusion that differing degrees of professionalism will everywhere produce comparable differences in police treatment of juveniles is unwarranted. Whatever the qualifications, however, the fundamental choice remains and no program of training or reorganization can escape it.

HERBERT JACOB

Wage Garnishment and Bankruptcy Proceedings in Four Wisconsin Cities

In recent years political scientists have focused most of their political participation research on the acts of voting, of running for political office, of campaigning for others, and of expressing demands through interest groups. These acts constitute in the main our image of political man. This characterization emphasizes his roles as a demander of governmental services and as a controller of public officials. Another role, intertwined with these, as a consumer of governmental services or as a user of public facilities, has not been much studied by us.[1]

The citizen's role as a consumer of governmental services has become more prominent with the growth of the welfare state. Many needs, if satisfied at all, were previously met mostly by private rather than public means. Education was provided mostly by private academies and denominational colleges. Private philanthropy furnished most welfare and health

The research on which this paper is based was supported by a grant from the Walter E. Meyer Research Institute of Law and by funds granted to the Institute for Research on Poverty at the University of Wisconsin by the Office of Economic Opportunity pursuant to the provisions of the Economic Opportunity Act of 1964. Donald Pienkos, Mary Ann Allin, and William Fisher aided considerably in the processing of the data. The conclusions are the sole responsibility of the author.
[1] Almond and Verba discuss one phase of the citizen's role as a consumer when they examine the concept of subject competence. Gabriel Almond and Sidney Verba, *The Civic Culture* (Princeton: Princeton University Press, 1963) pp. 214–257.

197

services. Insurance was supplied entirely by private enterprise and fraternal organizations.

Today these services are predominantly provided by the government. In addition, new public services are designed to meet a broad variety of needs for almost every citizen. Consequently, the public now consumes public services more frequently than ever before. For many citizens their principal contact with government is not through the suffrage but through public agencies which service their needs. Thus in 1960, there were 22.7 million people who obtained a large part or most of their income from government pensions; another 6.3 million received various kinds of public assistance payments. In addition, an estimated 26 million parents had children attending public schools or colleges. In these categories alone, 55 million people, or half the adult population, were consuming public services which were central to their life style.

Among the oldest agencies providing services for the general public are courts of law. They provide forums for airing private and public conflicts and for settling them according to established custom and law. Even such a traditional service has increased relevance to contemporary Americans. The provision of more public services and the extension of statutory law to many fields previously governed by private custom have made it possible to bring more issues to courts for settlement. More permissive attitudes toward divorce, the greater frequency of injuries from auto accidents, and higher urban crime rates have increased citizen contact with the courts. Available evidence suggests that the volume of litigation is now higher in most parts of the country than it was in the past.[2]

[2] See for example the following statistics and cases filed:

	Year	Cases per 100,000 pop.
Iowa District Court Cases Commenced	1956	845
	1965	1062
California Superior Court	1951–1952	2117
	1961–1962	2163
New Jersey Law Division and County Court Civil Cases	1949–1950	229
	1959–1960	332
Connecticut Superior Court and Court of Common Pleas	1929–1930	981
	1959–1960	995
Only Michigan Showed a Decline: All Trial Courts	1932	1370
	1960	1241

Sources. Iowa: Judicial Department Statistician, 1965 Annual Report Relating to Trial Courts of the State of Iowa, pp. 15 and 18.

California: Judicial Council of California, 19th Biennial Report, 1962, p. 143.

New Jersey: Annual Report of the Administrative Director of Courts, 1961–62, Supp. p. 5.

Connecticut: 17th Report of the Judicial Council of Connecticut, March, 1961, pp. 20–21.

Michigan: Office of the Court Administrator, Annual Report and Judicial Statistics for 1961, p. 80.

Citizen contact with the courts is likely to be as significant as contact with most agencies. In criminal matters the courts may deprive men of their liberty for long periods of time and brand them as convicts. In other cases the courts do more than provide a forum and the rules for settling private conflict or challenging the action of another government agency. Courts lend governmental power to the victor of a suit and enable him to force the loser to abide by the court decision. Many private conflicts are thus transferred to the public domain. Litigants may contend about private affairs but at the end both winner and loser find that the government has become intimately involved in their dispute.

The political consequences of the consumption of court services by various segments of the public are not at all obvious and have not been subjected to systematic empirical analysis. A few dramatic cases have changed the course of public policy[3] but most cases involve the application of existing policy.[4]

Court application of existing norms is likely to have several kinds of politically relevant consequences. The systematic application of norms may produce social and economic results that generate demands for legislative action to change the norms or the rules of their application. This is most likely to occur when the norms originated under different social or economic circumstances, but courts continue to enforce them without significant changes.[5] Court actions also frequently strike at the core of people's personal behavior, their life-style, or fortune. Such contact with government about personally significant matters is likely to color people's impressions of their government. Insofar as their impressions remain distinctive to the courts, they may bear little relevance to their evaluation of the political regime in general. If their impression of the courts colors their preception of government or the political regime in general, such contacts with the courts may become highly significant elements in the generation of support for the regime or alienation from it.

In this paper we will concern ourselves only with the conditions under

---

[3] A few such dramatic decisions by the Supreme Court have been examined. See the many studies of the aftermath of *Brown vs. Board of Education* and Frank Sorauf, "Zorach vs. Clausen: The Impact of Supreme Court Decision," *American Political Science Review*, **53**, 777–91 (1959).

[4] See H. Jacob, *Justice in America* (Boston: Little, Brown & Co., 1965), pp. 17–36 for this distinction and Joel Grossman, "Judicial Policy Making," paper presented in 1966 Midwest Conference of Political Scientists (mimeographed) for a critique.

[5] Some examples are: (1) The workmen's compensation laws which arose after attempted application of the fellow-servant rule to industrial accidents where the rule was inappropriate; (2) 19th century divorce laws in New York, the application of which finally provoked new legislation in 1966; and (3) current dissatisfaction with contributory negligence rules applied to auto accidents when the facts are usually too complicated for a just application of the rule.

which different groups avail themselves of legal remedies. In subsequent reports of this research project, the impact of involvement in litigation and its political significance will be explored.

A definitive answer to the general question of the conditions under which judicial action becomes relevant to potential litigants cannot be given at this stage of our knowledge. Since court actions may involve a broad array of conflicts and potential litigants, each type of conflict requires separate examination. Considerable attention has recently been given to comparable questions in the sphere of criminal law[6] and in the use of litigation to settle disputes arising from automobile accidents.[7] I have limited my research to the area of creditor-debtor relations and, within that, to the use of court action to collect delinquent debts by attaching the debtor's wages and the use of several judicial proceedings by the debtor to delay immediate payment or avoid payment altogether.

DEBTOR-CREDITOR CONFLICTS

Debtor-creditor conflicts arise out of a variety of situations. They almost always involve a refusal by the debtor to pay his creditor what the latter thinks is due him. Such a refusal may be based on the debtor's feeling that he has been cheated or that the creditor did not fully live up to his end of the bargain. A refusal may also arise from a sudden inability to repay because of unemployment, ill health, a family emergency, or pressure from other creditors to repay them first. Sometimes, debtors fail to repay simply because they forget or no longer want to repay. In each of these cases, the probability of conflict between creditor and debtor is high because most creditors will pursue the debtor until he has repaid as fully as they can reasonably expect. Their judgment of reasonable repayment may be based on a number of factors; the most important, however, is likely to be the cost of obtaining repayment judged against the amount of probable payment. When the cost of obtaining payment rises far above the amount owed, most creditors will write off the loan as uncollectable.

The range of individuals involved in these conflicts is immense in the kind of credit economy the United States has. Almost everyone in retail business, in service enterprises, and in the professions extends credit; in

[6] See Jerome Skolnick, *Justice Without Trial* (New York: Wiley, 1966); Wayne R. Lafave, *Arrest* (Boston: Little, Brown & Co., 1965); and Donald Newman, *Conviction* (Boston: Little, Brown & Co., 1966).
[7] Roger B. Hunting and Gloria S. Neuwirth, *Who Sues in New York City* (New York: Columbia University Press, 1962).

addition we have a multimillion dollar consumer finance industry. Thus, among creditors in such conflicts, we find finance companies, department stores, service stations, television repair shop proprietors, landlords, hospitals, doctors, and even lawyers.

With credit cards in almost every adult's wallet, the range of individuals involved as debtors in these conflicts is even wider. The only persons systematically excluded are the very poor who are without attachable incomes and who are unlikely to be extended credit.[8] Not every debtor is equally likely, however, to be involved in a credit dispute which will reach the courts.

A great variety of private actions may be undertaken in efforts to collect debts. When these fail, the courts stand ready to help creditors through a number of legal remedies. If an article was purchased through a conditional sales contract, the creditor may repossess it, sell it, and then collect from the debtor the difference between what it brought in the sale and what he still owes. Such proceedings frequently leave debtors owing substantial amounts for items they no longer possess because the resale value of many items bought on credit is less than the amount owed at the time of repossession. Creditors may also go to court to obtain a judgment against the debtor; such a judgment constitutes an official statement of the debt and authorizes the creditor to obtain the sheriff's assistance in collecting the amount due him. The sheriff may inspect the debtor's possessions and seize for sale any articles which are neither exempt from execution under the state's laws nor serving as security for another debt. When—as in many cases—the debtor owns no goods which may be seized to satisfy the judgment, creditors in many states have a final judicial remedy: they may attach the debtor's wages through a wage garnishment.[9] A summons is sent to the debtor's employer who is obligated to report whether he owes the debtor any wages; if he does, he must send those wages to court. The debtor may recover a token amount for living expenses, but the bulk of the funds so caught are used to satisfy the debt. A creditor may garnishee his debtor's wages as often as necessary to obtain full payment.

Debtors, in turn, also have a number of extralegal and legal remedies to which they may turn. In a country where there is free movement and different laws in each state regarding creditor-debtor relations, the easiest

---

[8] But note the marginal credit exposed by David Caplovitz, *The Poor Pay More* (New York: Free Press, 1963).

[9] A recent summary of wage garnishment laws may be found in George Brunn, "Wage Garnishment in California: A Study and Recommendations," *California Law Review*, **53**, 1222–27, 1250–53 (1965).

thing for a debtor to do may be to move away from where he incurred his debts.[10] Alternatively, he may defend his nonpayment in the judgment suit although that is expensive and rarely successful. Increasingly, debtors make use of a much more successful legal maneuver—promising repayment through a court-approved amortization plan. Court-approved amortization plans may be available under state law (as they are in Wisconsin) or through Chapter 13 of the federal bankruptcy statute. Under the Wisconsin statute,[11] a debtor earning less than $7500 per year may agree to repay his debts in full within a two-year period. During this interim, the debtor is protected from wage garnishments and other court actions seeking to collect the debts listed in the amortization plan. Debts not listed and new debts may be collected as before through judgments or wage garnishments. Amortization under Chapter 13 of the Bankruptcy Act[12] allows the debtor to repay over a three-year period; during this time, interest accumulation is stopped and all creditor actions against the debtor are prohibited. New debts as well as old ones may be included in the repayment plan although the debtor may incur new debts only with the approval of a court-appointed trustee. Chapter 13 plans may also provide for partial payment in satisfaction of the debt.

Bankruptcy provides the final legal escape for the debtor. Bankruptcy is equally available under federal law to the business and nonbusiness debtor. A debtor need not be penniless; he need only have debts which he cannot pay as they fall due. Under bankruptcy proceedings, the debtor makes available to the court all nonexempt assets that he possesses for repayment to his creditors. The definition of exempt assets depends upon state law; in Wisconsin they include the equity in the individual's home up to $5000 or $1000 in savings accounts, life insurance up to $5000, equity in a car up to $1000, one television set and radio, personal clothing, jewelry up to $400, beds and stoves used by the family, other furniture up to a value of $200, one firearm up to a value of $50, plus assorted miscellany such as a Bible, pew, cemetery lot, library, sewing machine, and tools of trade.[13] Consequently, almost all nonbusiness bankruptcies turn out to involve no assets that can be distributed to creditors. After this has been established, the federal court (through a special official, the Referee in Bankruptcy who handles these cases in place of the District Judge) discharges the debts of the bankrupt; he is therefore no longer

---

[10] The extent of debtor mobility is indicated by the proportion of debtors in our study who had moved away within 12 months of court action involving them. This proportion ranged from 11% in Kenosha to 21% in Green Bay.

[11] Wis. Stat. 128.21

[12] The statutory provisions for bankruptcy and Chapter 13 proceedings are found in 11 U.S.C. 1-1103.

[13] Wis. Stat. 128.21

legally obligated to repay. The single limitation to this remedy for debtors is that they may exercise it only once every six years.

CONDITIONING FACTORS

Only a tiny proportion of all credit transactions turn into conflicts between creditor and debtor and only a small proportion of those eventually are brought to court. It is therefore important to identify the conditions under which some people seek to invoke governmental sanctions in their efforts to collect or evade their debts. Five sets of conditions are readily identifiable.

First is the need of people to use the courts to resolve their conflicts. Creditors will not use the courts unless losses from credit transactions are high enough to reduce profits and cannot be recouped through extra-legal means. Debtors will not seek refuge in court remedies unless their debts become burdensome because of creditor pressure. The threshhold for creditors and debtors is different. For creditors it varies with the organizational structure of the firm and its sensitivity to customer relations. For debtors it depends upon their financial and psychological resources, their information about court remedies, and the size of their debt.[14]

A second conditioning factor is the availability of court action. In some communities, court action is unlikely because no court is sitting in the town—all actions must be started in a distant town making litigation inconvenient as well as expensive. Other factors, however, also make legal remedies variably available. Some courts are more stringent about requiring representation by lawyers than others; in some towns, attorneys are more readily available for collection work than in others; in some, attorneys charge more for collection work than in others. Finally, the cost of litigation varies from town to town, because local judges interpret state laws differently regarding fees, and local practices evade some of the more costly items.

Third is the ability of potential litigants to use remedies that are available. They need to know about them; they need the requisite financial resources; they must be convinced that court action is really appropriate in their situation; they need to be free of the psychological restraint of shame and the social restraint of retaliation. Different members of a com-

---

[14] "The Consumption of Governmental Services: Usage of Wage Garnishment and Bankruptcy Proceedings in Four Wisconsin Cities," Paper prepared for delivery at the 1966 Annual Meeting of the American Political Science Association (mimeographed); a fuller analysis will appear in a monograph to be published by Rand McNally in 1969.

munity are likely to vary considerably in their ability to use available remedies.

The socioeconomic conditions prevailing in an area is a fourth factor. A recession (even if slight and local) following a period of prosperity when credit was freely extended is likely to be more productive of creditor-debtor conflicts than a continued period of prosperity or a long recession. Likewise, the type of economy an area has is likely to be important. Subsistence economies (either in rural or urban slums) are not likely to involve much consumer credit. Factory workers whose employment or earnings are erratic are more likely to be caught in credit difficulties than are white collar workers whose employment is steadier and whose wages, although lower, are also more regular. The availability of credit is also significant; those living in small towns or in cities with few banks and lending institutions may find it more difficult to obtain credit and may also find themselves less frequently tempted to borrow than those living among a plethora of lending institutions and constantly being bombarded by an invitation to borrow.

Finally, the "civic" or "public" culture of a community is likely to be important. Some communities may be more conservative in lending policies than others; in some, large blocs of citizens may refrain from borrowing, either because they are older and not used to it or because they come from ethnic groups which are unaccustomed to living on credit. Alternatively, some communities are composed of groups who borrow heavily. In some communities, resort to court action comes easily without cultural constraints; in others, it involves a morally and culturally difficult decision. Thus court action may be much more frequent in some communities than others.

In this paper I shall discuss only the last variable. The others are investigated elsewhere.[15]

THE RESEARCH DESIGN

The research on which this paper is based was designed to obtain an empirical description of the conditions under which wage garnishments, court-supervised amortization plans and bankruptcy proceedings were used. Four middle-sized cities in Wisconsin were chosen as research sites. The four were chosen because they had been the sites for previous studies into their community power structures and therefore extensive background

[15] *Ibid.*

materials were available for them. Since all the cities examined were located in one state, the available legal remedies were held constant. In each city, court files were searched for wage garnishments and bankruptcy cases. In one city, Green Bay, where the number of such proceedings was small, the entire population of such cases for a 12 month period was recorded; in the others (Kenosha, Madison, and Racine) a systematic random sample was taken for a similar period. After checking for current addresses, a subsample of the remaining names was taken and hour-long interviews were obtained with debtors whose wages had been garnished, who had gone through Chapter 13 proceedings, or who had declared themselves bankrupt. A total of 454 interviews with debtors were obtained. In addition, a random sample of creditors and employers were sent questionnaires. Finally, personal interviews were held with selected attorneys, creditors, collection agents, and court officials. In addition to these interview and questionnaire materials, information available on the court records was recorded and used as a check against the survey data and a supplement to it.

A second phase of the research (not reported here) involved the use of a state-wide survey to provide a control group against which to measure behavior and attitudes of debtors.

FINDINGS

Neither creditors nor debtors use the courts to the same extent in the four cities we studied. Instead, gross variations in court usage are evident when we count the number of garnishment and bankruptcy actions initiated in a 12 month period (see Table 1).

Data for other court cases in these areas are also available, although

TABLE 1   GARNISHMENT AND BANKRUPTCY ACTIONS FOR 12 MONTH PERIOD: GROSS TOTALS AND ACTIONS PER POPULATION BY CITY

	Madison	Racine	Kenosha	Green Bay
Total Garnishments per	2860	2740	813	130
1000 pop.	22.6	30.7	12.0	2.1
Bankruptcies	112	100	63	32
Chapter 13's	37	18	5	7
Total BK-13 per				
1000 pop.	1.2	1.3	1.0	0.6

TABLE 2 CIVIL AND CRIMINAL CASES IN FOUR COUNTIES, JULY 1, 1964–JULY 1, 1965

	Dane (Madison)	Racine	Kenosha	Brown (Green Bay)
Civil Cases				
County and circuit court	1691	817	1203	803
Small Claims Court (minus garnishment)	4203	3024	743	233
All civil cases per 1000 pop.	18.9	21.3	7.4	1.9
Criminal cases (excluding ordinance)	3195	1020	867	722
All criminal cases per 1000 pop.	15.8	7.2	8.6	5.8

Source: Table 1, 1965 Annual Report of the Wisconsin Judicial Council Table 10. 1965 Annual Report of the Wisconsin Judicial Council.

only on a county not a city basis.[16] The frequency with which all civil cases except garnishments are initiated in the four counties (See Table 2) follows the same pattern as the garnishment and bankruptcy rates. The highest rate occurs in Racine with Madison, Kenosha, and Green Bay following. Criminal case rates do not follow the same pattern except that Brown County (Green Bay) ranks low there as it does in civil litigation. The distinctive pattern of criminal rates seems to reflect a different set of causal factors as we might expect, since criminal cases are initiated by public officials while civil cases are initiated by private citizens for the most part.

Some of the recent efforts to explain such differences in the output of governmental agencies have sought to explain them in terms of social, economic, and (where appropriate) partisan characteristics of the area.[17]

[16] The proportion of the city population to county population is not strikingly different for the four cities, except for Kenosha: The per cent of the county population in the city in 1960 was 57.2% for Madison, 59% for Racine, 67% for Kenosha, and 50.2% for Green Bay. The proportions of the population classified as urban in these counties were even more similar: Dane (Madison) 75%, Racine 73%, Kenosha 72%, and Brown (Green Bay) 80%. Consequently, the court data presented here should not reflect biases due to significant differences in size of urban population.

[17] *Cf.* the chapters by Richard Dawson and James Robinson, Robert Friedman, and Robert H. Salisbury in Herbert Jacob and Kenneth N. Vines, *Politics in the*

TABLE 3 SELECTED SOCIOECONOMIC INDICATORS FOR THE FOUR CITIES

	Madison	Racine	Kenosha	Green Bay
Population, 1960	126,706	89,144	67,889	62,888
Per cent nonwhite, 1960	1.9	5.4	0.5	0.4
Per cent with income between $3000 and $10,000, 1960	68.9	71.6	69.1	76.1
Per cent with high school or more education, 1960	65.3	40.6	36.8	46.3
Per cent migrants from different county, 1960	28.1	11.5	13.4	13.0
Per cent of labor force in manufacturing, 1960	15.1	49.0	50.6	26.9
Per cent in one unit structures, 1960	55.9	63.2	61.7	67.3
Per cent dwelling units owner occupied, 1960	54.2	65.0	65.7	65.0
Number of manufacturing establishments with more than 20 employees, 1958	45	82	37	42
Per capita retail sales, 1958	$1520	$1345	$1199	$1892
Per capita bank demand deposits, 1960	$931	$724	$558	$742

Source: Bureau of the Census, County-City Data Book, 1962, pp. 566–574.

Our four cities do vary considerably on important socioeconomic indicators as is shown in Table 3. Madison is the largest city. The high education and mobility of its population reflect the impact of the state university on Madison; the presence of the university and of the state government offices also depresses the per cent of the population employed by manufacturing firms. Racine, Kenosha, and Green Bay are much more similar but not identical. Green Bay has only a small portion of its labor force engaged in manufacturing; it has a relatively high retail sales volume. Kenosha, the home of American Motors, has the highest proportion of

American States (Boston: Little, Brown & Co., 1965); Richard I. Hofferbert, "The Relation between Public Policy and Some Structural and Environmental Variables in the American States," *American Political Science Review,* **60,** 73–82 (1966); Robert R. Alford and Harry M. Scoble. "Political and Socio-economic Characteristics of American Cities," *1965 Municipal Yearbook* (Chicago: International City Managers' Association, 1965).

its labor force in manufacturing but low retail sales volume and low per capita bank deposits.

These differences, however, do not help us explain the varying litigation rates. Madison's high garnishment and bankruptcy rate is incongruent with the high proportion of its population which is well educated, in white collar rather than blue collar occupations, and with large bank deposits. Green Bay's low garnishment and bankruptcy rate is all the more surprising with its relatively high proportion of people in the income bracket most prone to these difficulties—the $3000 to $10,000 range; likewise, its high retail sales volume and relatively high bank deposit rate would indicate a garnishment and bankruptcy rate like Racine's. Instead, Green Bay's garnishment and bankruptcy rate is half of Racine's.

Other studies have also shown that cities[18] and states[19] which have similar socioeconomic characteristics nevertheless have different patterns of political behavior. This may be because they have reached their presently similar position by different developmental paths and acquired distinctive styles of public action and evaluations of public activity. There are aspects of a city's culture which are likely to permeate its decision-making processes and also affect the likelihood of citizens bringing cases to court.

The relevant cultural characteristics of our four cities have been suggested by Alford and Scoble's examination of their decision-making processes. Alford and Scoble[20] hypothesize that these cities represent four cultures: (1) *traditional conservatism* (Green Bay) where "government is seen as essentially passive, as a caretaker of law and order, not as an active instrument either for social goals or private goals"; (2) *traditional liberalism* (Kenosha) where "the bargaining process may even extend to traditional services. . . ."; (3) *modern conservatism* (Racine) where "government is seen as legitimately active, but furthering private economic interests which are regarded as in the long range public interest"; and (4) *modern liberalism* (Madison) where "a high level of political involvement . . . may itself exacerbate conflicts . . ." Alford and Scoble

[18] See for example: Oliver Williams and Charles Adrian, *Four Cities* (Philadelphia: University of Pennsylvania Press, 1963) and Robert E. Agger, Daniel Goldrich, and Bert E. Swanson, *The Rulers and the Ruled* (New York: Wiley, 1964).

[19] All studies examining expenditures by states show a considerable residual remaining after socioeconomic characteristics are entered into the regression equations. See the studies cited in footnote 17.

[20] Robert R. Alford and Harry M. Scoble, "Urban Political Cultures," paper delivered at the meeting of the American Sociological Association, September, 1964 (mimeographed), pp. 1–5. Quoted by permission of the authors. Elsewhere (in a yet unpublished draft of a book) they define traditionalism in terms of low professionalization with consequent informality in decision-making and highly specialized leadership groups.

concentrate their analysis on the liberal-conservative dimensions of the four cities because they focus on the kinds of public decisions made and the range of participation in the decision-making process. The data in Tables 1 and 2 suggest, however, that the cities cluster on the traditional-modern dimension when we examine civil litigation rates. Thus it appears that Green Bay and Kenosha share low litigation rates because of their more traditional public cultures while Racine and Madison share higher litigation rates because of their more modern public culture.

What are the possible linkages between political culture and the civil litigation rates? Civil litigation requires decisions by individual plaintiffs and their lawyers to transfer a private dispute into the public domain of the judiciary. Of course, the propensity to litigate depends on the expected pay-off; few people will sue if the chances of winning judgment are small or if there are no funds to recover. But holding that consideration constant, we may hypothesize that in some cultures, commonly held values become social constraints against litigating; in the absence of such values, people are more likely to sue. Where a high level of government activity is valued and where publicity is not particularly feared because social relationships are at a bureaucratic rather than personal level, conflicts are more likely to ripen into litigation than where these conditions do not exist. Both a high valuation of government activity and bureaucratized relationships are associated with a modern rather than traditional culture. Data from our interviews indicate that these cultural conditions in fact exist in the four cities and are associated with the variation in litigation rates.

Taking a debtor to court in order to collect a loan is a highly impersonal proceeding involving the use of public officials as intermediaries (the sheriff serves the papers) and arbiters. Interviews with attorneys indicate that in Green Bay and Kenosha, creditors depend more on private means than do their counterparts in the modern cities, Madison and Racine. Green Bay and Kenosha creditors make more use of personal contacts, telephone calls to the debtor or informal arrangements with employers. In addition, fewer creditors in Green Bay and Kenosha thought that eliminating garnishment would make a difference to their operations. Moreover, in Green Bay where the garnishment rate is lowest, attorneys and creditors often asserted that the city was small enough for everyone to know everyone else, making court action unnecessary. In fact, Green Bay has only 5000 fewer inhabitants than Kenosha (where no respondents mentioned the intimacy of the town) and is only one third smaller than Racine with its high garnishment rate. Nevertheless, the exaggeration of its small size is significant, since the perception of Green Bay as a small town fits our characterization of its culture as traditional.

Further, the lower garnishment rate also fits Alford's description of these traditional cities as ones in which the business elite does not look upon government as an instrument to obtain its private objectives. If those businessmen who extend credit take their cue from the leaders of the business community, we would expect that they would not be as ready to use the courts to collect their accounts as their counterparts in cities where the business community frequently invokes government power to attain its objectives. Passive government and informal bargaining, typical of many public situations in Green Bay and Kenosha, also typify the debt collection process more frequently than in Madison and Racine. In Madison and Racine creditor-debtor conflicts, like public disputes, more frequently reach official government agencies (in this case, the courts) for formal adjudication.

The traditionalism we have spoken of spills over to the legal culture. The way in which attorneys handle garnishment cases in Green Bay is one indication of it. In Green Bay most of the attorneys who handle garnishment cases send letters to the debtors prior to initiating court action, even though cases referred to attorneys have already been extensively worked over by the creditor's collection department or a collection agency. Attorneys in no other city took this precaution or were as concerned to avoid formal court action. In the other cities, most attorneys reported that when a file was given them they immediately filed court action unless they knew that no one else had tried to collect. The distaste of the Green Bay attorneys for court action appears to reflect a more traditional, as well as conservative, attitude toward litigation. Attorneys in the other cities reflect more the modern view of litigation as a legitimate instrument to meet their objectives.

Alford's characterization of these four cities also leads us to look for differences in the degree to which people in these cities are fiscally traditional. In the two traditional cities, we might expect that even the debtor population would show greater distrust of finance companies than in the other two cities and that they would prefer, both in words and deed, such a respectable source of money as banks. In addition, we might expect the debtors of the two cities to express more conservatism in favoring borrowing for particular purposes than their counterparts in the "modern" cities.

These inferences are only partially supported by our data, as Table 4 indicates. Only in the traditional and conservative city (Green Bay) do debtors exhibit all the characteristics we expected: they are most approving of banks and least approving of finance companies as sources of loans; they actually used banks the most and finance companies the least when they borrowed money. They were the most conservative in

TABLE 4 MEASURES OF FISCAL CONSERVATISM AMONG DEBTORS IN THE
FOUR CITIES: PER CENT OF DEBTORS IN EACH
CITY INDICATING . . .

	Madison	Racine	Kenosha	Green Bay
Best source of loans				
Banks	47.9	48.9	43.5	54.4
Finance company	2.1	5.3	4.6	1.5
Worst source of loans				
Finance company	46.5	44.0	45.8	61.8
Largest source of credit				
Finance company	43.4	63.8	50.5	33.3
Banks	23.1	13.8	16.5	30.4
Knows interest rate for all or some loans	70.9	42.2	54.2	76.8
Approved borrowing for a large range of items	37.1	37.0	34.5	30.0

approving borrowing for a wide range of items. They most frequently could indicate the interest rates they thought they were paying for their loans. The pattern in the other cities is not so clear. Debtors in Kenosha are like those in Green Bay only in evaluation of borrowing; in their attitudes toward finance companies and banks, in their borrowing behavior, and in their knowledge of interest rates, Kenosha debtors are more like those in their sister city, Racine. On the other hand, debtors in Madison are more like those in Green Bay except in their approval of borrowing.

These inconsistencies warn us not to explain all differences by the political culture of these cities. The differences in Chapter 13 rates, for instance, are more clearly related to differences in perceived accessibility to Chapter 13 trustees as the result of advice given debtors by attorneys. Where Chapter 13 proceedings are most frequent (Madison and Racine), lawyers mention this alternative more frequently, thus giving the debtors a real choice between it and bankruptcy. In the other cities, lawyers rarely speak about Chapter 13 to their bankruptcy clients; since few clients hear of Chapter 13 from other sources, fewer debtors in those cities used Chapter 13 proceedings.

The reason that more attorneys "push" Chapter 13 in some cities than in others is related to the pressure they feel from the Referee in Bankruptcy and the accessibility of a trustee. In Madison, the Referee was a strong advocate of Chapter 13 and the trustee was exceptionally vigorous in making his services available. Racine's higher Chapter 13 rate (in contrast to Kenosha's, where the same Referee is operating) reflects the fact that the trustee for those cases was a fellow Racine attorney, readily available on the telephone for consultation. Kenosha attorneys also had to use the Racine trustee, but they hesitated to incur the long-distance charge for a call to neighboring Racine. In addition, they felt that Chapter 13 cases were beyond their control; they expressed a strong preference for amortization under state law (although it provided less protection for the debtor), because they could maintain control over the proceedings and because they could keep in close contact with the debtor who might bring them higher fees when he brought an accident case or a divorce case. Green Bay, Racine, and Madison attorneys almost never used state amortization proceedings and did not speak of them as a lure to attract clients in better paying cases. In this instance, a characteristic of the legal structure of the communities plays an important role in the frequency of debtor actions.

With four cities, it is statistically impossible to estimate how much of the variation is explained by the political culture, by the legal structure, and by what appear to be accidental variations. Four cases are sufficient to discern variation without establishing the distribution of these variations. Nevertheless, it is significant that the wide variations we discovered should exist among four cities which (although not identical) are much alike in their size, economic base, and social structure. If our analysis is correct, variation in the use of courts by debt collectors and by debt evaders is closely related to rather subtle cultural and normative factors, when gross socioeconomic indicators are held relatively constant as they were in this study. Our analysis does not indicate whether still larger differences in court usage might occur if we were to examine cities with grossly different socioeconomic characteristics. However, if large metropolitan areas are more like Madison than Green Bay in possessing a modern political culture with impersonal and bureaucratized relationships, we would expect that court usage would be still greater than in Madison and Racine.

CONCLUSIONS

This exploration of court usage in wage garnishment and consumer bankruptcy actions has not shown any of the usual political linkages between governmental action and private demands. This is because the usual linkages are missing. If we were to look for partisan biases of the judges (or referees), for evidence of other attitudinal biases in their decisions, for the linkage between the judicial selection process and court decisions, for the role of other political activities in this process, we would come away convinced that this is a totally nonpolitical process. Only an examination of patronage would evidence the existence of the usual political processes at work, since the referees may appoint at will Chapter 13 trustees and trustees for all bankrupt estates. However, Chapter 13 cases do not generate a great deal of revenue for the trustees and most bankruptcies by consumers involve no assets or very nominal assets so that the lawyers appointed to these cases benefit little from their appointment. In the usual *partisan* sense, the processes we have been examining are indeed nonpolitical.

Nevertheless, I would urge that such a view is erroneous. Garnishment and bankruptcy cases involve invocation of governmental power for private ends. The rules for these proceedings (as provided by the state legislatures and by Congress) invite invocation of governmental power to all who meet quite minimal conditions. The fact that only a few of all those eligible to use these proceedings do so is significant to the political system. In the first instance, it brings to the courts far fewer actions than might otherwise be the case. Secondly, it has consequences that may not have been intended by the authors of the laws which make court usage possible or which, when apparent, are unacceptable to concerned political leaders. Wage garnishments are used to buoy the socially least desirable form of consumer credit, that extended by finance companies. Bankruptcy is in fact used only by some of the debtors who might benefit from its provisions. The law is applied unevenly, even though the variation is unintended by legislator and judge alike. Third, the variation in court usage is likely to result in deepening disaffection among the disadvantaged.[21] Those most likely to be disaffected, the debtors who are socially not well integrated, perceive the courts only as an instrument of the creditor not as a source of relief from their own troubles. Where garnishments are more common or more concentrated among a particular segment of the population, they may therefore become a significant source of political alienation.

The political process I have been describing is beyond doubt quite different from the electoral or legislative political processes which political scientists ordinarily describe. However, it may bear a significant resemblance to a portion of the administrative process which is becoming increasingly important: the conditions of usage of governmental agencies

---

[21] This problem will be the focus of later reports of this research. Data about political efficacy were collected in the debtor interviews. However, see the speculation about the association between garnishments and the 1965 riots in the Watts area of Los Angeles: Ralph Lee Smith, "Saga of the Little Green Pigs," *The Reporter,* **35,** 39–42 (1966).

or consumption of government-provided services. The degree to which governmental action penetrates into the general social system depends to a great extent on the kinds of considerations we have examined wherever usage is voluntary rather than compulsory. In some administrative agencies the same kind of problem arises as in use of court services; this is true of the use of agricultural extension services, the use of counselling and educational services by the poor, and the use of higher-education facilities by the young—to name a few. None of these, however, involves the dramatic use of governmental coercive power over others for private objectives as does use of the courts.

As this study indicates, governmental power can be invoked through much more routine ways than campaigning in elections. This invocation is based on different objectives, and varies according to different characteristics of the invoker, than the ordinary use of governmental power through partisan means.

PART V

<hr>

Politics, Health, and Welfare

DONALD B. ROSENTHAL

ROBERT L. CRAIN

Structure and Values in Local Political Systems: The Case of Fluoridation Decisions

Emulating developments in the study of national political systems, recent treatments of local politics have moved from an emphasis on the politics and government of specific cities to the development of concepts to permit comparative analysis of the structure and political style of American municipalities.

Williams and Adrian, for example, classify four middle-sized Michigan cities according to their differing "local political values."[1] Such values are manifested in the conceptions held by citizens and decision-makers of the purposes their local government should serve. These may range from

Reprinted with modifications from the *Journal of Politics,* XXVIII (February, 1966), pp. 169–196. The research on which the present article is based was made possible by a grant from the United States Public Health Service to National Analysts, Inc., of Philadelphia. We wish to thank Aaron J. Spector of National Analyst for his cooperation throughout the many phases of the study. Our special thanks also go to Elihu Katz for the major part he played in the larger survey; and to him, John Crittenden, Morris Davis, and James Q. Wilson for comments on an earlier draft of the present paper.
[1] Oliver P. Williams and Charles R. Adrian, *Four Cities: A Study in Comparative Policy Making* (Philadelphia: University of Pennsylvania Press, 1963), esp. pp. 21–39.

217

simply operating a "caretaker" government maintaining certain traditional community services to acting as an "arbiter" among competing groups and interests in the city.

Similarly, Banfield and Wilson distinguish among voters on the basis of the "ethos" which they exhibit. Citizens adhering to a "Protestant ethos and middle-class political style" which they call "public-regarding" are said to act differently from those having an ethos derived from immigrant and working-class values ("private-regarding"). This raises the possibility that communities with large middle-class and native populations will tend to pursue those ends which appear to benefit all citizens and reject seemingly "self-interested" demands.[2]

On the basis of research in a number of Florida council-manager cities, Kammerer and her colleagues draw a distinction between communities which have monopolistic and competitive "styles" of politics. They relate these differences to such factors as demographic qualities in the municipalities themselves.[3]

In a related undertaking, Agger, Goldrich, and Swanson differentiate "power structures" and "regimes" at the local level. The former term points to the distribution of power in the community and the ideology of the political leadership; the latter to the "rules of the game" or values prevalent throughout the system. They then proceed to develop typologies for both of these dimensions.[4]

Whatever the specific shortcomings of each of these approaches, they have yielded suggestive answers to the question which *should be* central to any study of local decision-making: What are the factors which account for the observed differences among local political systems in handling demands made upon them?[5]

---

[2] James Q. Wilson and Edward C. Banfield, "Public-Regardingness as a Value Premise in Voting Behavior," *American Political Science Review,* LVIII (December, 1964), pp. 876–887. The translation of these personal values into a community ethos raises some complex problems. See Raymond E. Wolfinger and John Osgood Field, "Political Ethos and the Structure of City Government," *American Political Science Review,* LX (June, 1966), pp. 306–326, and an exchange between Wilson and Wolfinger in the December, 1966, issue.

[3] For some of their ideas see Gladys M. Kammerer, Charles D. Farris, John M. DeGrove, and Alfred B. Clubock, *City Managers in Politics* (Gainesville: University of Florida Press, 1962); and Kammerer and DeGrove, "Urban Leadership During Change," *Annals of the American Academy,* CCCLIII (May, 1964), pp. 95–106. Kammerer and her co-workers are talking only about one form of government—the council-manager system—in one state: Florida. The same limitation operates in another study where variations within one form of government in one state—the manager system in California—were investigated by Eugene C. Lee, *The Politics of Non-Partisanship* (Berkeley: University of California Press, 1960).

[4] Robert E. Agger, Daniel Goldrich, and Bert E. Swanson, *The Rulers and the Ruled* (New York: Wiley, 1964), esp. pp. 40–51, 69–124.

[5] By "local political systems" we mean throughout the legally described but com-

Obviously, all of these works point to the importance of *values* in accounting for variations in the decision-making process. In addition, we may find considerable support in the same literature for the suggestion that these values are expressed in different forms of government and in the acceptance of political practices like nonpartisanship and at-large elections. However, this relationship is rarely spelled out. Williams and Adrian indicate that a connection exists between "local political values" and political structure, but they do not make a systematic linkage except in a brief discussion of the built-in conflict between the manager plan and minimal "caretaker" governments.[6] Agger, Goldrich, and Swanson, on the other hand, suggest that the structure of decision-making and the local value system may operate independently. For example, they argue, the "regime" may undergo considerable change over time, while the power structure remains the same.[7]

We will proceed on the assumption that political forms are an expression of community values. It is almost impossible to establish over time, however (once political practices have become legitimated), what the direction of causalty actually is. Rather, it must simply be recognized that once it is established, a particular political structure may permit certain kinds of demands to become more salient and more legitimate than another structure would. As Charles Adrian has put it:

Structural arrangements do have an effect on government, but they neither guarantee good government nor prevent it. The forms are important because they affect the pattern of influence of various groups upon policy-making. The

---

munity-influencing decision-making structures which exist in a given municipality. The social context of such structures is of considerable importance as our discussion indicates. Our usage follows the model developed by David Easton in his "An Approach to the Analysis of Political Systems," *World Politics*, IX (April, 1957), pp. 383–400. Neither Easton nor any of those writers drawing upon his work has systematically attempted to apply his model to the local level, although Agger, Goldrich, and Swanson make some effort in this direction. Admittedly, there are many problems involved in such an application. Membership in a local political system in the United States involves very little commitment to a consensual system of values allegedly brought into play where the national political system is involved. For present purposes, however, it is sufficient to consider local decision-making as occurring within subordinate or truncated political systems where the values of the locality may only be significant for those who choose to be influenced by them.

[6] Williams and Adrian, pp. 282–287.

[7] Their discussion of "regime" resembles our consideration of the role of values. Their "power" structures, however, bear little relationship to a traditional consideration of formal political structures. Indeed, the amount of attention given to forms of government is meager; where structural factors are brought in, the major concern is with parties, but even then the emphasis is not on distinguishing communities on the basis of the nature of their partisan structures.

specific structure in any given case helps to establish behavior patterns and attitudes toward power that definitely affect the process whereby decisions are made.[8]

Thus, we are suggesting that electoral practices and forms of government (the "political structure" of a community) express (or cause) varying political values. This assumption is exceedingly important in developing a method of studying local political decision-making which will permit large-scale comparison, rather than the small-scale study which is part of the current stage of comparative analysis.

FLUORIDATION AS A TEST VARIABLE

As a test of the importance of political structure for differentiating decisional results, we turned to one of the most controversial subjects in the experience of many American cities: fluoridation.[9]

Since shortly after World War II, the subject of fluoridation has received considerable attention from social scientists. Because of the extremist tone of much of the opposition to this proposal to reduce tooth decay by adding minute quantities of fluoride to municipal water supplies, a striking amount of space has been given to the subject both in popular magazines and in scholarly journals with particular emphasis being placed on the heated referenda battles which have taken place.

For the most part, however, these studies have stressed the social and psychological processes at work in American cities. This has been done through analyses of *public* reaction to fluoridation proposals (by sampling public opinion or by looking at the results of voting behavior on referenda) or through investigations of the attitudes and actions of the issue partisans (opponents and proponents). With the single exception of an unpublished case study by Ravenna Helson and Donald R. Matthews,[10] political scien-

[8] Charles R. Adrian, *Governing Urban America,* 2nd ed. (New York: McGraw-Hill, 1961), p. 197.

[9] It is problematic how much controversy arises on any given issue discussed in the power structure literature. One reason that nonpolitical "influentials" may figure so prominently in decisions such as those discussed by Floyd Hunter and, more recently, by Robert Presthus in his *Men at the Top* (New York: Oxford University Press, 1964) may be due to the small number of issues on which there are meaningful differences in ideology between the nongovernmental "elites" and the "politicians," on the one hand, and the "elites" and the "public" on the other. Even Robert Dahl, whose approach is considerably at variance with that of Hunter, does not choose for discussion "issues" which are essentially at controversy. For a more sophisticated treatment of local decision-making which does take controversy and ideology into consideration, see the Agger-Goldrich-Swanson volume.

[10] Ravenna Helson and Donald R. Matthews, "The Northampton Fluoridation Referendum: A Case Study of Local Politics and Voting Behavior," unpublished paper, Smith College, 1959.

tists have given only passing attention to the fluoridation phenomenon and no one has tried to develop an understanding of the political systems within which these peculiarly volatile decisions have been made.[11]

In addition, studies of fluoridation controversies have been limited to the analysis of only one city or at most a small number of cities. The research reported here was designed to complement these other studies with an analysis which was (1) frankly political in approach, (2) based on a large number of cases.

Questionnaires were mailed to the local public health officer, the city clerk, and the publisher of the largest newspaper in each of 1,186 cities; 1,051 of these were cities of between 10,000 and 500,000 population which (1) did not have fluoride naturally present, and (2) were the primary consumers of their water supply. (This excludes many suburbs.) The remaining 75 cities were cities of between 5,000 and 10,000 population which had held referenda. These cities bias the sample of course, but their presence does not affect the direction or magnitude of any of the correlation reported here. Where they do have an effect, they are excluded. Other data were taken from the census and from the *Municipal Yearbook.*[12]

Response rates varied from 35 to 57 per cent for the three questionnaires. (Since a small number of cities—less than a fifth of the cities in our sample—have never considered fluoridation at all, this deflates the response rates below their true figures.) In the analysis that follows, the numbers of cases vary sharply, depending upon which, if any, of the questionnaires are used in each table. (One table, which requires responses to all three questionnaires for each city, has a very small number of cases and must be used gingerly for this reason.)

We realize, of course, that a complete analysis of political values cannot depend upon an analysis of only one issue, and we hope that other writers will undertake similar studies of other types of decisions.

From the responses to our questionnaires, we conclude that fluoridation has the following political properties: (1) It has almost unanimous sup-

[11] Students of community power structure have been concerned about distinguishing among communities on the basis of the locus of decision-making, i.e., whether a particular decision was made within the formal government structure or by some nongovernmental elite either for subsequent private associational performance or for governmental ratification. In this connection, on any political continuum of decisions made at the local level, fluoridation must be treated as one issue handled through *public* mechanisms rather than *private* ones.

[12] A complete discussion of the methodology and findings of the study reported here may be found in Robert L. Crain, Elihu Katz, and Donald B. Rosenthal, *The Fluoridation Decision: Community Structure and Innovation* (unpublished report to the United States Public Health Service: National Analysts, Inc., Philadelphia, 1964). A monograph covering this material is forthcoming, *The Politics of Community Conflict* (Bobbs-Merrill, 1968).

port among the elite. (2) Very few members of the elite have a strong interest in seeing it adopted. (3) The opposition comes from an organized minority, usually including chiropractors, Christian Scientists, natural food faddists, and in a few cases, the radical right. (4) The public at large is uninformed but is cautious because of the possible side-effects of any medical innovation. (5) The medical profession is virtually unanimous in support of fluoridation.

While each of these statements requires a slight qualification, the general implications seem clear to us: Fluoridation is an issue which can be adopted most easily by cities in which the political and civic elite's mild support is sufficient to offset the virulent opposition of a minority of ordinary citizens and only when it is possible to keep the level of public discussion within bounds so that the argument does not escalate into a confused debate involving large numbers of voters.

FORM OF GOVERNMENT AND THE FLUORIDATION DECISION

Since we are interested in governmental decision-making in this paper, our main variable is *not* whether the city adopted fluoridation or not but how the city council decided the issue. The council has three options: to adopt fluoridation (*administrative adoption*); to hold a *referendum,* which occasionally may be required if citizens take advantage of provisions for local initiative; or to reject flouridation without a public vote (*no action*). We are not here concerned with whether the referendum leads to adoption or rejection of fluoridation.

In Table 1, we pause to look at the sixteen cities which are too large to be included in our study. A single factor—whether the city is *legally* nonpartisan in its council elections or not—tells us a good deal about the actions taken by that city in regard to fluoridation. By 1965 *all* of the partisan cities had adopted fluoridation; the nonpartisan cities were divided among the various alternatives for action on fluoridation. If we had a more sensitive measure of true partisanship, we might be able to account for Chicago and Cleveland, where strong partisan organizations exist behind nonpartisan facades. Similarly, late-adopting New York City probably has a weaker and more fluid party structure than the other partisan cities. Cincinnati, with its traditions of nonpartisan "good government" under influential managers has held three vicious referenda campaigns on fluoridation. That the subject has come up several times is a tribute to the persistence of "good government" forces.

The present analysis is concerned with reporting the relationship between governmental structure and actions taken on fluoridation in smaller and middle-sized cities. It should be recognized that fluoridation is in some ways a "hard case" with which to test variations in community styles because it has had a rather lack-luster history under almost any conditions. In absolute terms, the differences we are actually describing are rather small. Of all the cities which have acted on fluoridation, only 35 per cent have adopted it administratively. Similarly, most referenda have led to fluoridation's defeat.

TABLE 1   PARTISANSHIP AND FLUORIDATION OUTCOME FOR
SIXTEEN LARGE CITIES: AS OF JANUARY 1965

Outcome	Partisan	Nonpartisan
Administrative adoption	Baltimore Buffalo Philadelphia Pittsburgh St. Louis New York	Chicago Cleveland Minneapolis
Held referenda		Cincinnati Milwaukee San Francisco
No action		Boston Detroit Los Angeles New Orleans

Note: Two cities are omitted. Houston has natural fluoridation, while Washington, D.C. has a Commission government appointed by Congress. (Fluoridation was adopted by Congress for the District of Columbia.)

The results given in Table 2, showing whether a city made an administrative decision to adopt fluoridation, decided to hold a referendum on it, or took no action, are consistent with our hypothesis that forms of government are associated with different styles in processing issues. Despite the small number of cases in some of the cells, the pattern is almost exactly the same in each region. In all regions, administrative adoptions are most frequent in council-manager cities, followed by partisan mayor-council cities. Referenda are more characteristic of nonpartisan systems, whether nonpartisan mayor-council or council-manager. In general, partisan mayor-council cities and commission towns are more likely to fall

TABLE 2 OUTCOME ON FLUORIDATION BY FORM OF GOVERNMENT
AND REGION (PER CENT)

		Mayor-Council		
Regions	Manager	Partisan	Non-partisan	Commission
Northeast				
Administrative adoption	24	15	9	5
Held referenda	15	5	25	11
No action	62	80	66	84
Total	101	100	100	100
N	(68)	(130)	(32)	(73)
South				
Administrative adoption	44	35	30	30
Held referenda	23	12	21	7
No action	33	52	48	63
Total	100	99	99	100
N	(127)	(40)	(33)	(43)
Midwest				
Administrative adoption	45	36	33	33
Held referenda	19	12	35	13
No action	35	52	32	54
Total	99	100	100	100
N	(110)	(94)	(72)	(39)
West				
Administrative adoption	9	—	8	—
Held referenda	26	—	24	—
No action	64	—	68	—
Total	99	—	100	—
N	(151)	(4)	(25)	(10)

Sources: Municipal Yearbook; Public Health Service Data.

in the "no action" category, with the commission government the most
likely in all three test cases.

INTERPRETATION

The fluoridation literature gives us few clues as to why such a relationship between governmental structure and decision-making appears in

Tables 1 and 2. Only one writer—James Coleman, in his important monograph, *Community Conflict*[13]—develops a theory linking community structure to the presence of conflict. He is not, however, concerned with governmental structure. (It seems possible to develop predictions about governmental structure from his paper, since he argues that conflict is more likely to occur when the government is remote from the public. As we shall see, our research indicates that this argument may be incorrect.)

In order to develop an explanation for Tables 1 and 2, we consider it desirable to create a typology using forms of government as *illustrative* of differing value systems and political structures. This does not mean we assert a clear relationship between form of government and every other aspect of the political system. Rather, what we are saying is that if it is valid to treat a municipality as a political system, then one of the ways in which the values of that system are expressed is through the political structure (including form of government) which that system uses.

First, for the purposes of analysis, communities which place a high value on wide-spread public participation in decision-making will be called *participative*. Such political systems are distinguished from others which restrict decision-making to formal political agencies. We shall say that these cities have adopted a *nonparticipative* political value or style.

Along with this normative or stylistic dimension, we shall also be concerned with one particular structural aspect of the local political system: the extent to which there is executive centralization. Robert Dahl has written of New Haven as an "executive-centered coalition"; we suggest that the degree to which power is centralized in the hands of the administrative and political leadership of a city is an important index of the decision-making *control* exercised. In the literature discussing differences between the "strong mayor" and "weak mayor" systems we find some support for this point, but there has been little systematic effort to trace the relationship between degree of formal governmental centralization and policy decisions. "Strength," of course, is a summary of many factors, but we will mainly be concerned with this one aspect.

The material in Figure 1a displays the possible relationships between executive centralization and the "participativeness" of the system. We take as a crude definition of participativeness the holding of many referenda in the municipality and as the definition of a strongly centralized

[13] James S. Coleman, *Community Conflict* (Glencoe: Free Press, 1957). For a consideration of the literature on fluoridation see Donald B. Rosenthal, "The Politics of Community Conflict," unpublished Ph.D. dissertation, Department of Political Science, University of Chicago, 1964, esp. pp. 89–179.

	Participativeness of System	
	Participative	Nonparticipative
Executive Centralization	(Many Referenda)	(Few Referenda)
Strong structure		
(Many administrative actions)	I	II
Weak structure		
(Few administrative adoptions)	III	IV

FIGURE 1a *System dimensions and predictions of community actions on fluoridation.*

decision-making structure the taking of actions at the administrative level. Our argument is that a centralized system would be *able* to act because there are fewer actors who must give their approval to the recommendations and because the centralization of authority would insulate the decision-makers from the antifluoridationists, either through the use of an appointed executive or the use of strong political parties controlling recruitment to office and protecting the incumbent from defeat.[14]

A city of Type I would be one in which action on proposed innovations is high at both levels: government is centralized, but the public participates heavily. Whether the city prefers to adopt administratively or to hold a referendum, it would be highly disposed to action of some kind. Because of a disposition to encourage public participation, the referendum is a readily accepted device for attempting to reach a decision. The presence of a strong executive, however, would encourage a decision to be taken through established decisional structures. As we shall suggest below, the actual form of government most closely resembling this type exhibits a marked split personality on the issue of fluoridation.

Cities of Type II, on the other hand, might be less readily activated in response to proposals for innovation but, having determined to act, would do so through established political channels, generally disdaining use of the referendum.

Cities of Type III, with little centralization and operating in a wide-open governmental milieu, would have difficulty in preventing the holding of referenda. The lack of a strong executive might mean an effort to shirk responsibility for action in an area of controversy—again a factor working for the holding of a referendum.

Finally, in Type IV cities there is neither an urge toward participation

---

[14] Given the general values of American society, we would further contend that the *ability* to act would lead to innovations under most contemporary conditions.

	Participative	Nonparticipative
	I	II
Strong structure	Nonpartisan	Partisan
	council-manager	mayor-council
	III	IV
Weak structure	Nonpartisan	Commission
	mayor-council	

FIGURE 1b   *System dimensions and related forms of government.*

built into the system nor a locus of executive responsibility. Such systems might be more readily inclined toward inaction.

While these types are "ideal," for present purposes they will be treated *as if* they were embodied in the four forms of government most common in the United States: the council-manager, partisan mayor-council ("strong mayor"), nonpartisan mayor-council ("weak mayor"), and the commission. The most interesting empirical question is how closely these "fit" the typology we have outlined. Figure 1b indicates the hypothesized relationship between real forms and the typology. Because published materials do not offer standardized measures of the degree of executive centralization in American cities, we have built into our analysis the dimension of partisanship as a part of "political structure."[15]

Gabriel Almond has suggested that one of the major functions performed by political parties in most political systems is the aggregation of demands. This function reduces the pressures playing directly on the decision-making machinery. Such an "insulation" effect, we hypothesize, occurs at the local level by the presence of strong political parties. To the extent that decision-makers are secure in their positions, they are able to act on the basis of criteria which might not be immediately acceptable to certain vocal segments of the public.[16] For the purpose of argument, therefore, we assume that partisan mayor-council systems are

[15] There are other types of government, of course; the partisan council-manager form, for example, is probably the strongest structure and is probably less open to participation than the nonpartisan council-manager city. The town meeting is an even weaker and more participative system than is the nonpartisan mayor-council form. Thus these two forms could be added to cells II and III in Figure 1b. Our data do not permit us to discuss two other types, the partisan mayor-council system with a weak executive and the nonpartisan mayor-council system with a strong mayor.

[16] Gabriel Almond, "Introduction," in Gabriel A. Almond and James S. Coleman, eds., *The Politics of the Developing Areas* (Princeton: Princeton University Press, 1960), pp. 38–45. This point is further developed throughout Myron Weiner, *The Politics of Scarcity* (Chicago: University of Chicago Press, 1962).

more likely to have powerful mayors than nonpartisan mayor-council systems. Many of the former may appear to be formally weak mayor systems, but even where this is the case, the presence of a party system may help to overcome the formal weaknesses of the local executive, as it has done in Chicago.[17]

The city manager form of government, on the other hand, is accompanied in its classic version by nonpartisanship and a "political formula" which favors the referendum, thereby weakening formal political authority. However, the system is favorable to innovation (at least in theory) and the manager is vested with considerable administrative authority. Thus, we would expect the system to favor fluoridation and the manager to seek some action on the subject, but the participative bias of the system means that if any controversy is raised on fluoridation, the system will have considerable difficulty in resolving the issue. This flows from the "consensual" style which is built into manager systems, i.e., they have a capacity for acting where issues can be defined as administrative or noncontroversial, but situations of real political conflict strain the structure. When citizens are activated under these conditions, the lack of control over their participation may disrupt the system. It is partly for this reason that many matters may be referred to the public for decision or called from the hands of the government by public demands for a referendum.

In the remainder of this paper we will develop additional data to support the argument that these two variables—participativeness and executive centralization—are reasonable explanations of the correlation between the form of government and the fluoridation decision. We cannot, unfortunately, develop very pure measures nor can we disprove all the possible alternative explanations. However, we can show in a variety of ways that our data are consistent with our interpretation, though we must leave some questions open.

EXECUTIVE LEADERSHIP AND STRUCTURAL "STRENGTH"

The central finding of the study of fluoridation decision-making is that the position taken on fluoridation by the mayor virtually determines the outcome.[18] We have data on 496 decisions made in 362 cities, and (as

[17] For a portrait of the ways in which party organization can overcome the forces of decentralization see Edward C. Banfield, *Political Influence* (New York: Free Press, 1961).
[18] Our findings with respect to the influence of the system's executive are reported in Donald B. Rosenthal and Robert L. Crain, "Executive Leadership and Community Innovation: The Fluoridation Experience," *Urban Affairs Quarterly,* I (March, 1966), pp. 39–57.

Table 3 indicates) if the mayor does not support fluoridation, there is very little chance for its adoption. Indeed, the mayor's neutrality on fluoridation is almost as unfortunate for the proponents as is his outright opposition. On the other hand, if the mayor does support it, fluoridation has a better-than-even chance of adoption without a referendum.

TABLE 3   MAYOR'S POSITION AND GOVERNMENTAL
DECISION ON FLUORIDATION (PER CENT)

Governmental Decision	Mayor's Public Position		
	Supported	Neutral	Opposed
Adoption	54	5	6
Hold referendum	13	21	29
No action	33	74	65
Total	100	100	100
N	(258)	(189)	(49)

Note: The cases are decisions, rather than cities; the 496 decisions come from 362 different cities.

The close correlation between the mayor's position and the outcome may seem obvious; after all, he is the most important figure in the government. The finding is interesting, however, not only because it runs counter to the argument of some "power structure" theorists, but also because the mayor is by no means the only decision-maker on the issue—in fact, it is usually the city council which has formal responsibility for the decision. In many cities, furthermore, the mayor has little authority, holding a largely honorific post.

However we feel about the preceding table, there is no obvious reason why the mayor should have a great deal of influence upon the public vote in a referendum on fluoridation, as indicated in Table 4. Yet, we see that the mayor's public position has a great deal of effect and in much the same way as it did on the governmental decision. Fluoridation will rarely be approved unless the mayor publicly endorses it; it has about an even chance of adoption if he does endorse it.

There is some justification, then, for treating the mayor's stand as a cause of adoption, rather than a merely spuriously associated correlate. This suggests that we could fruitfully analyze the factors which cause one mayor to support fluoridation while another opposes it. We assume that there is enough slippage in the system to permit an element of free choice to operate, and we have shown elsewhere that in the case of city managers such factors as age, education, and professional mobility are associated with willingness to support fluoridation. However, in this paper we are concerned almost entirely with the way in which political structure constrains the political leader, and we shall not touch upon the effect of personality.

TABLE 4   MAYOR'S POSITION AND REFERENDUM DECISION
ON FLUORIDATION (PER CENT)

| | Mayor's Public Position | | |
	Supported	Neutral	Opposed
Decision			
Percentage of referenda leading to adoption[a]	46	11	7
N	(34)	(37)	(14)

[a] In some cases, fluoridation was endorsed by the public vote but never installed.

Let us first take a closer look at the effect of partisanship, with a somewhat better measure than simply the presence of party names on the ballot. Two questions were addressed to publishers in the cities we surveyed. "How important are political parties in the elections for city council and mayor?" and "Do these tend to be local or nationally affiliated parties?" We assume that the more "local" a party is, the less it will act as a strong Republican or Democratic party is assumed to do and the less local candidates will be able to depend upon coattail effects to guarantee their election. The local party is more likely to be fluid in its membership, better able to enter coalitions with other groups, and more vulnerable to reform or capture. In all these respects, the local party is "weaker" than the sort of organization we hoped the publisher would identify as "national" in orientation.

In Tables 5 and 6, we look not at the decision itself but at the health officer's report of whether the mayor publicly endorsed fluoridation in mayor-council cities. These tables illustrate a pattern which we anticipated: the more local politics is influenced by political parties and the more national the identification of the parties, the more favorable the mayor will be to fluoridation. The mayor's freedom to follow the lead of the elites and his bureaucracy reflects the "strength" of the political order. The few mayors actually opposed to fluoridation tend to come from cities where parties are locally based or elections nonpartisan. These results, furthermore, are not merely the consequence of the regional distribution of governmental forms; indeed, controlling for region *increases* the correlation. For example, if we remove the effects of region (by the techniques of standardization), we find that in mayor-council cities where the publisher considers parties to be very influential in elections, 69 per cent of the mayors favored fluoridation; in mayor-council cities where parties have no influence the percentage drops to 48 per cent. This is a difference

TABLE 5 PARTY INFLUENCE AND MAYOR'S STAND, IN MAYOR-COUNCIL CITIES (PER CENT)

Mayor's Stand	Very Influential	Moderately or Not Very Influential	Not at All
Favor	71	55	55
Oppose	5	4	8
No stand	24	41	37
Total	100	100	100
N	(41)	(56)	(62)

TABLE 6 NATIONAL AND LOCAL PARTIES AND MAYOR'S STAND, IN MAYOR-COUNCIL CITIES (PER CENT)

Mayor's Stand	National Parties	Local or Mixed Systems
Favor	71	51
Oppose	2	8
No stand	28	41
Total	101	100
N	(51)	(39)

of 21 percentage points, compared to the difference of 16 shown in Table 3. In a mayor-council system that demands leadership from the mayor, the chances appear greater that he will assume that policy-leadership on fluoridation under conditions of partisanship. In a nonpartisan system, we suggest, the mayor is much less likely to feel responsibility for upholding the claims for support made by the local bureaucracy.

The finding that mayors in partisan towns are more favorable to fluoridation than in nonpartisan towns is interesting not only in itself but because it was the thesis of the early reformers of local government that nonpartisanship would attract to office men likely to work for "good government" measures. While we are reluctant to classify mayors' attitudes toward good government by their stands on fluoridation alone, there does seem to be some support here for those who argue that nonpartisanship has not lived up to the promises made for it. Furthermore, the weakness of the nonpartisan mayor-council system in this respect cannot be compensated for. At least in the council-manager system, there is an alternate policy leader available.

Charles Adrian, drawing upon material from two city councils and two state legislatures, has proposed some general principles concerning the effects of nonpartisanship. He argues, for example, that nonpartisanship can attract the successful businessman into local government (as the system was intended to do), but that at the same time it encourages the perennial office-seekers or anyone who has had the opportunty to make his name familiar to the voters, even if it is merely because of a history of opposition.[19]

Following this reasoning, we might hypothesize that the nonpartisan office-holder would be more reluctant to support fluoridation than the partisan, since he must retain his office without party support, solely on the impression he has given the voters. It seems reasonable that he would want to avoid touching a sensitive nerve in even a small portion of the population by supporting fluoridation. On the other hand, a marginal candidate, a relatively unknown alderman in a large council, or a mayor whose powers are so limited that he is unable to get much attention from the newspapers—all might see opposing fluoridation as a possible way to gain attention. (Supporting fluoridation, on the other hand, does not seem to guarantee the electoral support of profluoridationists, most of whom are not strongly interested in the issue the way the "anti's" appear to be. In addition, supporting fluoridation is not an unconventional position and hence not as newsworthy.) In contrast, the partisan official might concern himself with the attitude of the party slating committee and the party's financial supporters toward him, for he might require such support to be renominated and reelected.[20] This might lead the official who is in a more tightly structured political role to avoid "irresponsible" *opposition,* though this is no assurance he would *favor* fluoridation.

Throughout our discussion of the relationship between political structure

---

[19] Charles R. Adrian, "Some General Characteristics of Non-Partisan Elections," *American Political Science Review,* XLVI (September, 1952), pp. 766–776.

[20] Given the large number of instances in which mayors do not support fluoridation, this consideration should not be overstated. It should be stressed that even in partisan towns the intraparty primary provides an opportunity for a candidate to use the support of the public to undermine the party organization. In broad terms, however, it remains a tautology that strong party organizations are able to exercise control over their members' actions, which is all we are saying here.

and fluoridation decisions, we have emphasized both the types of people who are recruited by the several systems and the kinds of restraints placed upon their behavior by the contexts in which they find themselves. The manager is frequently a professional with a commitment to the use of "principles" of administration. The partisan mayor is perhaps more "responsible"—not so much because of his personality or training but because of the system in which he is functioning. He is, therefore, less likely than a "loner" nonpartisan figure to take a radically negative stand on an issue like fluoridation. Indeed, he is more likely to take a positive position, because he has party support and because he is more likely to feel responsible for the operation of the system and hence want to act as a spokesman for the city health department. On the other hand, even the commitment of the nonpartisan mayor to "good government" may not give him enough strength to take a stand against the opinion of any sizable (or noisy) group.

If it is true that the differential allocation of power to the executive is a major variable in determining the actions taken by each city in dealing with fluoridation, we should be able to see it by comparing the positions of "weak" and "strong" mayors on fluoridation, using other measures of "weak" and "strong" than simple partisanship. The former should be less favorable to fluoridation than the latter. This distinction can be crudely measured by the mayor's length of time in office (arguing that a mayor's term of office is an indirect correlate of his independence and power) and by whether or not he has a veto over the council's decisions. Both indices were applied to 140 cities for which both the health officer and publisher had returned questionnaires. We find in Table 7 that mayors

TABLE 7 TERM OF OFFICE OF THE MAYOR AND HIS STAND

Term of Office	Per Cent of Mayors Favorable	N
1 year	36	11
2–3 years	59	75
4–5 years	70	54

with longer terms of office are noticeably more favorable, as predicted by the hypothesis. Short terms of office tend to occur in nonpartisan cities, so the relationship is as expected. The nature of the mayor's veto power, however, does not appear to be an important influence on mayoral action.

Additional material, much of it anecdotal, suggests that systems in which policy-making is concentrated in a few hands are better able to cope with the controversy which surrounds fluoridation. In particular, cities which must deal with an independent water supply (either privately owned or operated by a regional authority) are less likely to adopt fluoridation because the water supplier is an additional actor with a partial or complete veto to override. The fight to eliminate decentralization of existing city governments has been going on for a long time, but many governments still have a collection of appointive and elective officials only vaguely responsive to each other.[21]

Fluoridation is the kind of issue which is peculiarly vulnerable in such a decentralized decision-making structure. It is an issue with an excellent ability to attract headlines; at the same time, it is not typically an "important" issue to which civic groups and leading businessmen will flock. In many cities, it is easy for officials to ignore fluoridation as not worth the noise and the opposition liable to arise, for fluoridation is the kind of unpleasant issue in which opposition can generate headlines about "encroaching socialism" or "putting the bureaucrats in control of men's minds."

The more decentralized the system, the greater the number of decision-makers who must make the difficult decision in favor of fluoridation. Given this situation, opposition by one official can often stop adoption; decentralization would thus appear to have particularly unfortunate consequences for fluoridation. We can even argue that doubling the number of decision-makers will more than double the chance for opposition. Each of these leaders is competing for the scarcest of resources in the political game: public attention and personal influence. If the rewards of conformity must be divided many ways (and divided unequally at that), then the unsuccessful politician may find that he can reap more benefit from opposition than from support. If the opponent has a veto, he can often set a high price on his consent. Even if it is legally and institutionally possible to override the objections of the minority, to do so may require considerable expenditure in time, energy, or political capital. One of our findings is that fluoridation does not have many friends with this amount of power and the willingness to use it. The reader may recognize this argument as similar to the one made by Amos Hawley in his study of urban renewal

[21] For discussion of this problem see Banfield and Wilson, *City Politics* (Cambridge: Harvard University Press, 1963), especially Chaps. 6 and 8. The study by Helson and Matthews also gives an excellent example of how a system in which there are a great number of institutionalized participants complicates the decision-making process. This same sort of problem appears to be at work in the case of the Jacksonville, Florida, school system, where so many governmental actors have their fingers in the making of school budgets and school policy that at one point the system reached near collapse and lost its national accreditation.

TABLE 8   MEAN NUMBER OF ELECTIVE OFFICIALS FOR CITIES SURVEYED
AND FLUORIDATION ACTION IN FOUR STATES[a]

	State			
Action	Michigan	Ohio	North Carolina	Minnesota
Held referenda	10.1 (11)	14.6 (8)	8.8 (8)	12.0 (6)
Administrative adoption	9.9 (19)	12.8 (11)	8.7 (12)	11.0 (4)
No action	8.7 (9)	12.0 (36)	7.8 (9)	10.1 (9)

[a] The number of cities in each category is indicated in parentheses.

adoptions: cities with many influentials will have power more widely distributed and thus be less able to act.[22]

Thus, if too many cooks spoil the fluoridated broth, cities which have the most difficulty with fluoridation should be those with the largest number of elected officials. In a quick test of this rather extreme hypothesis, four large states from the Midwest and South were selected at random and the mean number of elected officials, as given by the United States Census of Governments, was compared against fluoridation action where the data were available. The result is shown in Table 8. In all four states, the rank order was identical. (There is only one chance in 216 of this occurring by chance.) The cities which held referenda had the largest number of officials. After this surprisingly easy confirmation of our hypothesis, however, we find that in every case cities which have adopted fluoridation administratively have more elected officials than cities which have not taken any positive action on fluoridation. The reasons for *not* acting, however, are so diverse that it is difficult to provide an explanation for this except to indicate that it runs contrary to our initial expectations.[23]

PARTICIPATIVENESS

We have suggested that local political values vary in the importance given to public participation and the attention paid to minority opinion.

[22] Amos Hawley, "Community Power and Urban Renewal Success," *American Journal of Sociology*, LXVIII (January, 1963), pp. 422–431. Hawley implies that middle-class communities have more difficulty instituting urban renewal plans because they are likely to allow more scope for participation in the making of the decision under the style which we have described as "participative."

[23] While the data merits only a note, we should report that in three of the four states the cities which *won* referenda tended to be cities which had few elected officials. Of course, the number of cases involved is very small.

One measure, as previously indicated, is the number of referenda that the system holds. The commission system rarely uses fluoridation referenda, but it also rarely adopts fluoridation, so that the lack of referenda may also mean that this particular form of government is incapable of any action on fluoridation. Of the remaining systems, it is the nonpartisan systems which hold the bulk of the referenda, and number of referenda held is a key factor in the failure of fluoridation.

Table 9 demonstrates the differences among the various political systems based on information provided by the health officer in each city. We asked him whether the possibility of holding a referendum was ever discussed in the city and whether one was actually held. The table demonstrates that partisan cities were less likely to discuss the possibility of holding a referendum and even if one was discussed, they were relatively unlikely to hold one. Nonpartisan mayor-council cities, on the other hand, were most likely to consider holding a referendum and most likely to hold one, with the manager cities a close second in both cases.

The city with a participative style is likely to develop political practices over time which revolve around the public meeting, the citizen's committee, and the referendum. Fortified by nonpartisanship, almost any group of citizens can try to find and support a candidate for office and his chances for election may be meaningful. In such a system any vocal group must be recognized as part of the legitimate opposition.

Sometimes this participative style works in favor of fluoridation. The manager system, for example, is most likely to have the proposal for fluoridation introduced by a civic group, rather than by the public health professionals. Thus, the greater willingness or ability of citizens to participate means that fluoridation will have more supporters as well as more

TABLE 9   VARIATION BETWEEN COMMUNITY "TALK" OF REFERENDUM BEING HELD AND ACTUAL HOLDING BY FORM OF GOVERNMENT

| | Form of Government | | | |
| | | Mayor-Council | | Com- |
	Manager	Partisan	Nonpartisan	mission
Was there ever talk of holding a referendum? (Per cent "yes")	46	35	54	47
N	(223)	(147)	(82)	(81)
Was one held? (Per cent "yes")	48	29	52	25
N	(104)	(49)	(44)	(36)

TABLE 10 PERCENTAGE OF CITIES RECONSIDERING INITIAL ACTION ON FLUORIDATION, BY TYPE OF GOVERNMENT, WITH EFFECT OF TYPE OF INITIAL ACTION REMOVED

| Initial Action on Fluoridation | Form of Government | | | |
| | | Mayor-Council | | Com- |
	Manager	Partisan	Nonpartisan	mission
Percentage reconsidering	36.7	29.6	40.7	41.1
Expected percentage based on distribution of initial actions taken	34.2	36.0	35.6	41.1
Difference, between lines 1 and 2: the net effect of form regardless of type of initial action	+2.5	−6.4	+5.1	0.0
N	(215)	(142)	(76)	(73)

opponents. In addition, if the participative system rejects fluoridation, it is more likely to have the issue come up again, so that fluoridation gets more chances to be adopted. Unfortunately for fluoridation, the same is true once it is adopted; participative cities are more likely to cancel the program later. The top line of Table 10 demonstrates that the more participative systems are more likely to reconsider their decisions. The high percentage of commission governments reconsidering is a result of their tendency to take "no action," which is the easiest kind of decision to reconsider. Therefore in Table 10, we have controlled for the effect of the initial decision by standardization. In line two, we have presented the expected percentage of reconsideration, if the forms did not vary in the probability of reconsidering a particular kind of action but only varied in the type of initial action they took.[24] Then line three (which is the difference between lines one and two) shows the net effect of form

[24] If the above explanation is unclear, note we have simply computed for each form of government:

$$\varepsilon_a = \sum_{x=1}^{x=4} \left(\begin{array}{c}\text{proportion of all cities} \\ \text{reconsidering action "}x\text{"}\end{array}\right) \left(\begin{array}{c}\text{percentage of type "}a\text{" cities} \\ \text{taking initial action "}x\text{"}\end{array}\right)$$

where ε_a is the expected percentage reconsidering for cities with type "a" form of government, and $x = 1, 2, 3, 4$ represents four types of initial action (adopt, win referenda, lose referenda, no action).

of government on the chances for reconsideration, independent of the effect of initial type of action. As anticipated, the manager and nonpartisan mayor-council cities are more likely to reconsider than would be expected, the partisan mayor-council cities the least likely.

It is worth noting that citizen participation has a rather surprising effect on reconsideration. Not only are those cities which involve the citizens in decision-making least able to make a stable decision, but in addition, adoption by referendum is less likely to "stick" than adoption by administrative fiat (the data are not shown). We would think that resorting to a majority vote of all the eligible citizens would be a way to settle an issue once and for all. Such is not the case.

We would also expect that if the conventional wisdom about democracy is valid, then the community which over the years encourages its citizens to participate in decision-making would educate its citizens to take more "rational" actions. In this case, we know from the work of Gamson[25] and others that the voters reject fluoridation primarily because they are afraid of medical side effects and not because of any principled opposition to "socialism." We know from our own work that they are more likely to vote "yes" if fluoridation is endorsed by the mayor and the local medical organizations. We also know that Americans generally have high respect for medical science. Thus, it follows that the citizen's "rational" decision is to accept the advice of the near-unanimous medical profession. Now let us look at cities which have a tradition of citizen participation in referenda. Table 11 presents the referenda results in cities which have held over five referenda in the past twelve years compared to those which have held fewer.

At this point, what we find is no longer surprising. The cities with participative governments behave no more "rationally" than those which do not. In addition, they are less likely to be able to settle the issue with a single decision, and they are just as likely to have controversy even though these cities do not need controversy as a pretext for holding a referendum. Finally, they are not more likely to vote for fluoridation despite the fact that their mayor is more likely to support it. We cannot present the cross-tabulation controlling for mayor's stand since the reports are from different cities. However, it is logical that if we could, we would find that, "controlling" for mayor's position, the more experience a community had with referenda, the more likely it would be to reject fluoridation.

The reader may notice another rather complex interaction effect here. Nonpartisan cities have mayors who are less likely to support fluoridation. But in the table above, cities which hold many referenda are more likely

[25] William Gamson, "The Fluoridation Dialogue: Is It Ideological Politics?" *Public Opinion Quarterly*, XXVI (1962), pp. 526–537.

TABLE 11   BEHAVIOR OF HIGH AND LOW REFERENDA SYSTEMS:
REFERENDA CITIES ONLY

Referenda Systems	Cities Holding Five or More Referenda in Past Twelve Years	Cities Holding Less Than Five Referenda in Past Twelve Years
Per cent of referenda leading to adoption	26 (77)	27 (56)
Per cent of referenda actions "complex"[a]	39 (77)	21 (56)
Per cent of referenda hotly debated[b]	53 (34)	57 (30)
Per cent of mayors favorable	62 (26)	41 (22)

[a] "Complex" actions involve at least two actions: either two referenda, an administrative adoption preceding a referendum, or some other combination.
[b] Derived from a question to publisher: Compared to other issues with which you are familiar, would you say that fluoridation was: very calmly discussed; calmly discussed; warmly discussed; hotly debated?

to have mayors supporting fluoridation during a referendum. The answer to this puzzle is simple, but interesting. If a mayor in a partisan city supports fluoridation, it will be adopted administratively in most cases. The referendum is used when the administration wants to be neutral. In the nonpartisan city, however, the referendum must often be used by a mayor who supports fluoridation, since he lacks the power or party discipline to get it adopted administratively. We must qualify this argument, however, since the difference in percentages of mayors favorable is only significant at the 10 per cent level.

When we turn from the process of decision-making to the mechanics of the battle itself, we see some interesting differences in the extent of public participation in partisan and nonpartisan systems. Since nonpartisan systems hold more referenda, they should be expected to display more of the preconditions of high controversy. However, even when we compare *only* cities which held referenda, the nonpartisan mayor-council and manager systems are more likely to have held many public meetings. We asked health officers to indicate how many meetings were held on fluoridation. In nonpartisan mayor-council systems, 58 per cent of the cities re-

TABLE 12 POLITICAL EXPERIENCE OF OPPONENT LEADER AND OUTCOME, BY PARTY INFLUENCE (PER CENT)

Outcome	Parties Moderately or Very Influential		Parties Not Very or Not at All Influential	
	Opponent Experienced	Opponent Inexperienced	Opponent Experienced	Opponent Inexperienced
Administrative adoption	31	29	23	22
Held referendum	33	15	38	49
No action	36	55	40	29
Total	100	99	101	100
N	(36)	(65)	(48)	(96)

porting indicated that many meetings were held during a referenda campaign. In partisan mayor-council systems, however, only 47 per cent of the campaigns were marked by many meetings.

Some of the difference in political action can also be traced to the interaction between the system and the opposition. In the course of the study, the characteristics of persons who were regularly identified as leading proponents and opponents were analyzed. In general, we found that proponents were more likely to have a college education, a high-status occupation (especially that of physician or dentist), and to be well-known in the community (on the basis of the judgment of the publisher of the leading newspaper); in two cases out of three, however, they had no past political experience.

The opponent leader, typically, was older, was not a college graduate, and had a somewhat less high-status occupation (though not as distinct from the proponent leaders as we might have presupposed).[26] He was less well-known than the proponent and even less politically experienced. Out of the 269 opponents who were characterized, only 87 had some prior activity in political life.

Table 12 compares the relative effectiveness of politically experienced and inexperienced leaders in cities with "weak" and "strong" parties. There is no difference in administrative adoptions, contrary to what we might expect, but there are striking differences in the number of referenda held. The partisan systems, as we know, hold fewer referenda but when they do hold them the opponents are more likely to have political experience.

[26] For a further discussion of the activities of the proponents and opponents of fluoridation, see Crain, Katz, and Rosenthal, *op. cit.*

In contrast, the nonpartisan system seems to hold its referenda in response to the nonpolitical opponent. This again suggests the possibility that the referendum may serve different functions in the two systems: in the partisan system it may be used to relieve a stalemate within the government itself, whereas in the nonpartisan system it may reflect a situation where the structure is more vulnerable to the politically "unsocialized" opponents, i.e., the "radicals" of the community.

While we mentioned the partisan system as providing "insulation" from public participation, we did not specify any of the factors working towards this effect. One such element is the function performed by a minority party. When the minority party refuses to champion opposition to fluoridation, persons opposing fluoridation are left without any easy way to present their case. On most issues, the minority party may consider whether opposition will lose them support within the party, from "good government" elements, or from financial backers. These possibilities do not appear likely in the case of fluoridation. In addition, they may estimate that the antifluoridationists are not a lasting source of political backing. Table 13 supports this point by demonstrating the superiority of the two-party

TABLE 13 PARTY STRUCTURE, FORM OF GOVERNMENT, AND OUTCOME[a]
(PER CENT)

Outcome	Two-Party	Local Parties	Nonpartisan	One-Party
		Mayor-Council		
Administrative adoption	38	28	23	20
Held referenda	10	19	43	20
No action	52	53	34	60
Total	100	100	100	100
N	(63)	(32)	(33)	(30)
		Manager		
Administrative adoption	38	32	22	18
Held referenda	33	21	33	46
No action	29	47	44	36
Total	100	100	100	100
N	(21)	(34)	(90)	(11)

[a] The data are derived from the publishers' responses to the question: How many of the present members of the city council are supported by the local Republican organization? The local Democratic organization? Other local parties? (Space for three) Independent?

system as far as administrative action on fluoridation is concerned. For both mayor-council and manager forms of government, the two-party system is marked by the largest proportion of administrative adoptions (and, in the case of mayor-council cities, the smallest proportion of referenda held). Next best, in both cases, is the system of local parties where structural competition may still exist. The two nonpartisan systems lag far behind.

CONCLUSIONS

What we have found, we think, indicates that fluoridation has a better chance of consideration and adoption where the following conditions are met: there is a local political structure characterized by decision-making authority, centralized in a relatively strong executive like a manager or partisan mayor; the political structure provides the mechanisms through forms of government and strong parties which insulate mayors and managers from "irregular" pressures likely to arise on this issue; and, finally, that there is a low level of direct citizen participation both as a general rule and specifically on the fluoridation decision.

Broad popular participation, particularly in the absence of strong executive leadership and an institutionalized channel for confining the expression of opposition, spells defeat for fluoridation. We have elsewhere argued that it does so because fluoridation is a technical issue, the advantages of which are rather small from the citizen's point of view and equally minor from the viewpoint of the politicians as far as political capital is concerned.[27] On an issue of this sort the opposition can easily implant doubt. Doubt takes root and blossoms, the more the issue is discussed. For whatever reason the issue is raised for public discussion—whether it is because of a democratic debating tradition in the city's associations or a tradition of holding referenda—the opposition succeeds in arousing the citizenry to vote "no" and endless exhortations by proponents under the rubric of "educating" the public seem to fail, often rather badly.

Obviously we have not been able to produce "clean" measures for all the concepts we have put to use, but the very fact that some of the crude measures we did employ showed up relatively well indicates that there is considerable need for indices of community decision-making which are able to take into account structural factors related to the local polity and which probe the political values of a community.

[27] *Ibid.*

MARTHA DERTHICK

Intercity Differences in Administration of the Public Assistance Program: The Case of Massachusetts

Formally, there are no intercity differences in state programs of public assistance. The federal government, which pays more than half the cost of public assistance, prohibits them. Since it began giving assistance grants to the states in 1936, it has prescribed uniformity within each state as a condition of its aid.[1] For each category of federally aided assistance— Old Age Assistance, Medical Aid to the Aged, Aid to the Blind, Aid to the Permanently and Totally Disabled, and Aid to Families with De-

This research has been financed by the Joint Center for Urban Studies of MIT and Harvard.

[1] Uniformity *within* states but not *among* them. States may decide within broad limits who shall be eligible for assistance and how much they shall receive. As a result, there are substantial interstate differences in the content of public welfare programs, and these have recently drawn the attention of political scientists. See Richard E. Dawson and James A. Robinson, "The Politics of Welfare," Chap. 10 in Herbert Jacob and Kenneth N. Vines, eds., *Politics in the American States* (Boston: Little, Brown, 1965), Richard I. Hofferbert, "The Relation between Public Policy and Some Structural and Environmental Variables in the American States," *American Political Science Review*, LX, 73–82 (March, 1966), and G. David Garson, "Structural and Environmental Factors in State Welfare Policies," unpublished seminar paper, Harvard University. Federal administrators would prefer nationwide uniformity as well, but Congress has stood in their way.

pendent Children[2]—the Social Security Act requires the state to have a "plan," which in practice consists of the laws and regulations governing administration of that category. The plan must "be in effect in all subdivisions of the State, and, if administered by them, be mandatory upon them." The federal intent is elaborated in the Welfare Administration's *Handbook of Public Assistance Administration*:

> The purpose of the requirements relating to State-wide operation is to assure that the public assistance programs in which there is Federal financial participation, will be in effect throughout the State; . . . that the benefits of the programs will be equally available to all eligible persons; and that State policies, standards and methods will apply equally to persons in like situations wherever they may live.
> Where the program is administered by political subdivisions under State supervision, the State agency is responsible for assuring that the State program and State rules and regulations are in effect in all political subdivisions.

It is reasonable to suppose, however, that state and local practice may not conform perfectly to the federal intention, and that, informally, intra-state differences occur in public assistance administration. Federal supervision is carried out by small regional staffs that, at least in New England, spend little time observing the conduct of local administration. The Welfare Administration depends upon state welfare agencies to see that uniformity is achieved. It would be surprising if they did not find it difficult to write rules and enforce them in such a way as to assure that administrators in diverse local situations, confronted with diverse problems and cases, carry out the rules, respond to the problems, and handle the cases in precisely the same way.

This article reports on the extent to which uniformity exists in one state, Massachusetts. It has three purposes: (1) to identify differences among local welfare agencies in administrative outputs; (2) to isolate the extent to which such differences depend upon the agencies' use of discretion and can therefore be said to reflect differences in "leniency"— the use of discretion to increase expenditures for the benefit of the client; and (3) to explain why some agencies are more lenient than others, if they are, and whether differences in the use of discretion are related to the type of community in which the agency functions.

THE CASE OF MASSACHUSETTS

The language of the federal laws and regulations suggests that uniformity may be harder to achieve in some states than in others, depending on

---

[2] There is a residual category, general relief, that is not federally aided and to which the requirement of uniformity therefore does not apply.

whether the state administers public assistance itself or authorizes local subdivisions to do so. Twenty-eight of the states administer it themselves. All but one of the other 22 rely mainly on their counties. In the remaining one, Massachusetts, it is administered by cities and towns.[3]

The system of public assistance administration in Massachusetts is the most decentralized in the country. Legally, each of the 351 cities and towns runs an assistance program, although some of the smallest (in response to carrots and sticks from the state) have combined to form welfare districts, an arrangement that enables them to share staff without surrendering legal autonomy. The number of administrative units has thereby been reduced to 270. Most other states, whether administration is carried out by county governments or by the state through district offices, have fewer than 100 administrative subdivisions. Georgia, with 159 (county) units, is the next most decentralized.

The extreme decentralization of the Massachusetts administrative structure does not necessarily mean that administration is less uniform than in other states. The degree of uniformity depends on the content of state laws and regulations and on the effectiveness of the state welfare agency in enforcing them, factors that may be affected by the amount of decentralization but are not altogether determined by it. Among many other things, the interest of the governor and legislature in welfare programs, the competence and energy of state welfare commissioners, and the state's traditional division of roles and powers with its local governments, not only in public assistance but in other programs, may have an effect.

Administrative decentralization in Massachusetts is combined with a high degree of centralization in rule-making. The Massachusetts legislature (which is large, quite open to pressure from interested portions of the public, and—perhaps on that account—prolific) has produced a very detailed public assistance statute; although the detail is found mainly in the sections relating to old age assistance (which have been influenced by a lobby of the aged) and is much less characteristic of the sections covering aid to dependent children. Matters that are not covered by statute are covered in the Department of Public Welfare's *Public Assistance Policy Manual,* a detailed set of rules that is binding on the local agencies and serves as "the bible" for local caseworkers, each of whom has a copy. Thus, it is state laws and rules that the local worker is engaged in carrying out. Local agencies have no rule-making power except in general relief. Boston alone, with by far the largest caseload and most mature, elaborate administrative organization, has emended and clarified state rules with issuances of its own that are inserted in locally used copies of the *Manual.* In other places, written, locally prepared rules do not exist.

[3] As of 1968, this will no longer be true. In 1967 the legislature enacted a plan for state administration.

The state rules are particularly detailed and explicit with respect to the two basic administrative decisions involved in public assistance—who is eligible and how much money he shall receive. Since 1942 the state has issued a standard budget that is mandatory upon local agencies; it is designed to assure that persons in like circumstances, wherever they live, will receive identical amounts of money. The state rules are enforced by a staff divided among six district offices and consisting mainly of forty-six field representatives who routinely visit local agencies, sample case records, and observe administrative activities. The local agencies are free to exercise discretion only with respect to those few matters about which the state has been silent or explicitly permissive.

UNIFORMITY AND THE CATEGORIES

This article deals with intercity differences in administration of a single category, AFDC. Federal laws and rules prescribe uniformity within the categorical programs of aid but not among them. That is, elderly poor in like circumstances must be treated alike, dependent children in like circumstances must be treated alike, and so on, but there is no formal requirement of equity as between, say, the aged and dependent children.

The decision to examine only one category was based on convenience. The choice of AFDC was made for several reasons. Nationwide and in Massachusetts, it is the largest category.[4] It is also the most controversial, partly because it is growing rapidly and therefore seems to substantiate the fear that the assistance program creates dependency, and partly because it challenges prevailing mores. Most of the recipients are in homes broken by separation or divorce. About a fourth of the children are illegitimate. Some of the recipient mothers are immoral by any standard, and many more are widely thought to be. Critics of the program charge that it involves the public in approval and support of conduct that it does not approve and does not wish to support.

It seemed likely that if intercity differences exists in public assistance administration, they would appear in the AFDC program. This supposition rested on two grounds. One was that local resistance to state-prescribed rules would be greatest with respect to this category (and only strict adherence to the rules can produce uniformity). State rules tend to embody the values of professional social workers, who hold several of the top policy-making and supervisory jobs at the state level and are even more

---

[4] In December 1965, AFDC accounted for 4,456,995 of the 7,533,340 persons receiving public assistance nationwide and 103,002 of the 201,202 receiving it in Massachusetts.

dominant in the federal administration, which influences the content of the state rules. In recent years the rules have steadily changed to the benefit of the recipients. Local workers and administrators, who are almost completely unprofessionalized, often share the traditional, middle-class, moralistic values of the local community from which criticism of the AFDC program springs. Unlike professional social workers, many of them are not reluctant to pass judgement, which is sometimes but not invariably harsh, on the morality of welfare recipients, and some are frank in deploring the liberal trend of the rules, which they believe encourages dependence. The second supposition was that opportunities for the exercise of local discretion would be greater in AFDC than in other categories because it is governed less by statute and more by administrative rules than the others are. In Massachusetts, the low status of AFDC among public assistance categories is indicated not by the harshness of law but by the absence of it. The legislature, while passing many bills for the benefit of the aged, has been indifferent to AFDC. As a result, the program is governed mainly by provisions of the *Manual,* which—while technically binding on local agencies—is less awe-inspiring than a statute. Besides, in some important matters the rules have been explicity permissive. For example, the state has required local agencies to allow OAA recipients their choice of physicians, but, until 1966, the *Manual* merely suggested that AFDC recipients be given the same benefit.

In summary, it was supposed both that local agencies had more discretion in the administration of AFDC and that they were more strongly motivated to make use of what they had, and possibly a little more than they had. The logic of this supposition appeared increasingly dubious as research proceeded. To the extent that disapproval of AFDC exists among local administrators, it tends to produce uniformity rather than the reverse, by limiting the cash grant and other expenditures to a basic, state-prescribed minimum. Discretion, it turned out, tends to be exercised only positively: it takes the form of adding on to a minimum standard rather than shaving off. The restrictive agency is the one that fails to add on. There is probably more adding on, and thus more evidence of intercity differences, in OAA than in AFDC.

Most of the following analysis deals with the 29 cities and towns that had 100 or more AFDC cases as of July 1965.[5] It is necessary to concentrate on the largest places because most of the outputs to be compared

---

[5] The 29 range in size from Gloucester (25,789) to Boston (697,938). The others are Beverly, Brockton, Cambridge, Chelsea, Chicopee, Everett, Fall River, Fitchburg, Haverhill, Holyoke, Lawrence, Lowell, Lynn, Malden, Medford, New Bedford, Pittsfield, Quincy, Revere, Salem, Somerville, Springfield, Taunton, Waltham, Watertown, Weymouth, and Worcester. For convenience, I shall refer to all of them as "cities" although technically Watertown and Weymouth are towns.

are averages—e.g., average monthly cash payment per recipient—and where caseloads are small, the averages are vulnerable to distortion by a few deviant cases.

ADMINISTRATIVE OUTPUTS: HOW AND WHY THEY DIFFER

Differences among localities in administration of public assistance might occur in any of three types or stages of activity: (1) determination of eligibility, which yields a "recipient rate"; (2) determination of the monthly cash grant and other expenditures to or on behalf of the recipient, which yields "expenditure rates"; and (3) intangible treatment of agency clients, i.e., behavior that expresses the agency's attitude toward its clients without affecting its recipient or expenditure rates. This article will deal almost exclusively with the first two, and mainly with the second, because they are much more susceptible than the last to observation and objective analysis.

The Recipient Rate

Cities in Massachusetts differ considerably in their AFDC recipient rates—children receiving AFDC as a proportion of children under 18. Among the 29 for which data were assembled, the rates in 1960 ranged from .78 to 5.96 per cent. These variations might be explained by differences in the social and economic characteristics of the populations, such that the proportion of formally eligible persons varies, or by differences in the administrative behavior of welfare agencies. Either populations are dissimilar, or eligibility rules are applied differently to similar populations.

The first explanation is the more powerful of the two. Through correlation and multiple regression analysis, it was found that three socioeconomic variables—percentage of women separated and divorced, percentage of families with annual income below $3,000, and percentage nonwhite of persons under 18—together "explained" 62.4 per cent of the variation in AFDC rates.

Some portion of the "unexplained" percentage is probably accounted for by differences in the behavior of local welfare agencies. Two cities had recipient rates more than a third higher than they "should" have had—and eight had rates more than a third lower than they "should" have had—given their socioeconomic characteristics. The presumption is strong that welfare agencies in at least some of these cities were applying eligibility rules in a fashion either lenient or harsh by comparison with most other agencies, although it is possible that much of the unexplained variation may not be the result of agency behavior but of some unobserved socioeconomic characteristic of the population. Furthermore, knowledge of the availability of public assistance and willingness to apply for it may be greater in some places than others regardless of how the agency acts.

Expenditure Rates

Assistance expenditures are of two kinds: cash grants, which go directly to recipients, and vendor payments, which go to suppliers of goods and services. Most vendor payments are for medical purposes; they go to hospitals, physicians, dentists, and druggists. A small fraction of vendor payments goes to retailers and suppliers of nonmedical services.

In Massachusetts in 1965, AFDC expenditures were divided as follows:

Cash grants	86.7 per cent
Medical vendor payments	12.1 per cent
Nonmedical vendor payments	1.2 per cent

All these types of expenditure are governed by state rules, but either explicitly or by omission, the rules permit discretion in some matters. This makes it theoretically possible to use expenditure rates as indicators of the relative leniency of local agencies. For this purpose, medical vendor payments are the least satisfactory of the three, for the consumption of medical care by welfare recipients is subject to a wide variety of influences in addition to those that originate in the welfare agency. The cash grant is more satisfactory. Though it is determined for each case by the local agency's application of a state standard budget, the rules permit increases for "special needs." Thus, a lenient agency may make marginal additions to the standard allowance. Nonmedical vendor payments are the most satisfactory indicator, although diminished in importance by the fact that they are such a small fraction of total expenditures. They are used for "nonrecurring needs" such as the purchase of furniture, purchase and repair of appliances, and moving of household goods. The state permits such expenditures without requiring them or defining the circumstances in which they should be made. Thus they are wholly within the discretion of the local agency.

Local agencies report monthly expenditure data to the state department of public welfare. The data for AFDC in 1965 revealed variations among the 29 cities as shown in Table 1. Note that the table gives average expenditures *per recipient*. On the assumption that the average case consists of four recipients, one adult and three children, then the variation *per case* is four times what the table shows.

These data seem to indicate that federal and state intervention has

TABLE 1

	Cash Grant	Medical Vendor	Nonmedical Vendor
Average per recipient per month	$42.86	$7.16	$0.35
Range	$35.93–$51.86	$4.39–$10.32	$.01–$1.21
Range as percentage of average	83–121	61–144	3–345

not been very effective in producing uniformity in the administrative behavior of local welfare agencies, but it would be wrong to arrive so quickly at that conclusion.

For one thing, a comparison with expenditure data for general relief (the residual category of aid not subject to federal and state supervision) shows that intercity variation is much greater for general relief than for AFDC. Among the 29 cities, the average AFDC grant ranged from $35.93 to $51.86, the average deviation was $1.92, and the standard deviation $2.76; for general relief, the range was from $16.37 to $91.33, the average deviation was $16.08, and the standard deviation $20.60.

For another, the principle of uniformity does not necessarily imply that expenditure rates must be the same everywhere and for all three types of expenditure. There are many possible explanations for intercity differences in addition to the theory that agencies use their discretion differently. Intercity differences in the cost of shelter might affect the cash grant. If rents in any city exceed the allowance in the state standard budget, the local agency may, with state approval, grant increases to all or selected recipients. If all agencies responded to this rule in the same way, intercity differences in the cash grant would result—and be positively correlated with rent levels—but the principle of uniformity would not be violated. Differences in the capacity of local police departments and district courts to collect support from absent fathers may also contribute to variation in the cash grant. Intercity differences in hospital rates and in the prescription practices of doctors affect medical vendor payments. So may differences in the availability of medical facilities. Intercity differences in nonmedical vendor payments, most of which are for the purchase of furniture and appliances, might result from differences in rates of in-migration (on the supposition that in-migrants are more likely to be in need of these items). This list could be made longer.

Insofar as it was possible to do so, such explanations were tested statistically.[6] In general, the results were negative or inconclusive. The possibility remained that the differences resulted to a significant degree from differences in the welfare agencies' use of discretion. This possibility was explored through field research.

THE USES OF DISCRETION: TWO CASES

To determine whether outcomes do reflect differences in the use of discretion and to explore the reasons for such differences, two contrasting cases were selected for study. Those selected were cities that consistently ranked near the top or the bottom on all three of the expenditure rates explained above and on a fourth—total expenditure per recipient per month.

When the 29 cities were ranked according to the three rates, in most cases there was little correspondence among the ranks. In other words, a city that ranked high on cash grant per recipient did not necessarily rank high on medical vendor or nonmedical vendor payments. The correlation between cash grant and nonmedical vendor was only .37; between nonmedical vendor and medical vendor, .20; and between cash grant and medical vendor, .34. Of the 29 cities, only 10 were consistently in either the upper or lower half of the rankings. Among those 10, 2 were consistently in the highest quartile and 2 in the lowest. One city was chosen for study from each of these pairs on the assumption that one would prove to be lenient and one not.

City X ranked fourth in cash grant per recipient, third in medical vendor, seventh in nonmedical vendor, and third in total expenditure. City Y ranked 27th, 25th, 24th, and 27th, respectively. In 1960 X had an AFDC rate of 2.24, 12.7 per cent below the predicted rate of 2.52. Y's rate was 1.53, 87.3 per cent below the predicted rate of 2.87.

Both X and Y are medium-sized and, although they were not chosen for that reason, could be regarded as fairly typical of the 29. They are

---

[6] Expenditure rates for the 29 cities were correlated with selected socioeconomic variables—per cent nonwhite under 18, per cent foreign stock, per cent migrant, ratio of white-collar to manufacturing employees, median family income, median gross monthly rent, and per cent owner-occupancy. The highest correlations were between medical vendor payments and per cent owner-occupancy ($r = .30$), median gross monthly rent ($r = .32$), and per cent nonwhite ($r = -.31$). It is quite likely that these correlations reflect variations in the supply of and demand for medical services that exist independently of the actions or attitudes of local welfare agencies. All other r values were less than .3, and most were near zero. Given the fallibility of the expenditure rates as indicators of leniency, the correlations (no matter what their results) would be of little use in explaining why some agencies are more lenient than others.

old, poor manufacturing cities, historically dependent on a single industry but forced in the last few decades to adjust to the loss of much of it. Both have a low ratio of white-collar to manufacturing employees; indeed, X has the third lowest ratio among the 29. Like all Massachusetts cities, they have a high proportion (between a third and a half) of foreign stock, and like all except Boston, Springfield, and Cambridge, they have only a negligible proportion, less than 2 per cent, of nonwhites.

Of the two, Y is slightly larger and slightly poorer, yet as of 1965 it had fewer assistance cases in all of the federally aided categories except OAA. It had many more cases on general relief, a less generous program that entails less administrative inconvenience for the local agency and no federal and state surveillance. Apparently, some cases that in X would be aided under one of the federal categories are aided in Y through general relief, a further indication of the contrasting character of their administration. Table 2 shows the differences in the percentage distribution of cases among categories.

TABLE 2

	Number of Cases	Per Cent				
		OAA	MAA	DA	AFDC	GR
X	1634	42	25	9	14	9
Y	1552	46	20	9	11	14

Field research established that the differences between X and Y in expenditure rates did in fact reflect differences in the agencies' use of discretion. Precise comparisons are difficult because of Y's relative inaccessibility to research. It was impossible in Y to consult case records or observe the work of the agency, but qualified sources provided enough information to make rough comparisons possible.

In general, X freely uses its discretion to provide items above the minimum standard of expenditure whereas Y does not. Y buys furniture and appliances occasionally, normally from the Salvation Army or the Society of St. Vincent de Paul. X frequently makes such purchases, usually from a retailer. In 1965 X provided such items in 25 per cent of the 230 cases that received aid for six months or more; a total of $3,110 was spent. Y does not pay clients' moving expenses, with the possible exception of cases in which the move is caused by a public program. X routinely pays for one move a year, and more if the need is urgent or demonstrably desirable; in 1965 it spent $3,067 for moving, divided among 76 cases.

Neither agency makes a policy of paying clients' accumulated utility bills, a problem that plagues welfare agencies. However, Agency X in fact spent $1,323 for that purpose in 1965, mainly to prevent disconnections during cold weather. In Y clients who desire medical care must first obtain an authorization from the department. If they need transportation to hospitals or clinics, caseworkers are expected to take them in their own cars, for use of which the city pays a mileage allowance. Only if a caseworker is not available does the agency pay for other means of transportation. In X, clients need not obtain prior authorization for medical care, and if they cannot transport themselves to the place of such care, the agency pays taxi fare. In 1965 it spent $1,386 for that purpose. X increases rental allowances in numerous cases. Since the spring of 1966, too late to affect the expenditure figures used here, it has routinely granted $5 to $8 monthly increases to all cases in which they are warranted by the amount of actual expenditure. Y almost never increases the monthly rental allowance.[7]

None of these differences in agency practice, however, accounted fully for the difference in expenditure levels reported to the state. As it turned out, much of the *reported* difference in the cash grant ($45.39 in X, $40.19 in Y) was not an *actual* difference, but the cause of the reported difference was itself significant as an indicator of contrasting administrative practices.

Until October 1966, when the state issued a rule to assure uniform practice, local welfare agencies followed either of two procedures in calculating budgets of AFDC families to whom absent fathers had been ordered to provide support. In determining the amount of family resources (the cash grant equals need less resources), some agencies included the amount of the father's support payments and some did not. The former gave what was called a "deficit grant," the amount of need less the presumed contribution from the father. If, as often happened, the father failed to pay, the mother had to call or visit her caseworker and ask the agency to make up the deficit. This was the procedure used by Agency Y. The alternative, which was used by X, was to give a "full grant," the total amount needed. Fathers' payments were made directly to the city; if a

---

[7] This difference between the two agencies may be based on actual differences in rent paid by clients in the two cities. According to the 1960 census, the median gross monthly rent in X was $5 higher than in Y. A high proportion of Y's clients live in public housing, and many of the rest pay no more for rent than the standard budget allows. Whatever the precise differences between X and Y in this matter, it is clear that X's practice of granting rental increases makes it lenient by comparison with most other cities. The median rent in X in 1960 was well below the average median for the 29 cities, but as of 1966 X is one of only two cities that have put higher rental allowances into effect for AFDC.

father failed to pay, the mother did not have to appeal to the welfare department for an additional check in lieu of his. The state department of public welfare favored the latter practice as more humane, had consistently encouraged it, and finally, in 1966, required it.

Because X gave a full grant, the figures it reported for cash grants were larger than what it actually had to spend. X recovered $42,000 in 1965 from absent fathers, and its average cash grant—when calculated on the basis of net rather than gross expenditures—was $40.73, only $0.54 higher than Y's. Nevertheless, the reported difference between the two agencies remains significant as an indicator of differences in intangible treatment of recipients. The practice of giving a full grant is consistent with other evidence of X's leniency.

The differences between X and Y cannot be explained by differences in community values, at least as they are expressed by elected officials who might be supposed to have influence on the cities' welfare programs. Among public officials in both cities there is latent resentment against mothers who, as the aldermanic president in Y said, "are just raising children, not families," and against "frauds, cheats, and exploiters," in the words of the welfare board chairman in X. If popularly elected local governments had power to make welfare rules, there is little doubt that the rules would be more restrictive and that an effort would be made to draw distinctions between cases that are deserving according to the standards of the local community and those that are not. There is no reason to suppose that X would be more generous than Y, although—because there is presently no opportunity in either X or Y for community preferences to be expressed—it is hard to be sure about this. The important point is not that community officials' attitudes in both places appear to be similar but that in both places they are largely irrelevant.

In most Massachusetts cities, including X and Y, public assistance is formally the responsibility of a citizen board, which is usually appointed by the mayor or manager but may, as in Y, be elected by the aldermen. Before federal and state intervention robbed local governments of their rule-making function in poor relief and compelled them to hire civil service personnel to perform administration, local welfare boards were extremely powerful. They both made rules and carried them out. Now they do neither, and in most places they have become anachronisms. Except for serving informally as channels for complaints—by workers against the director or by clients against the workers—they have almost nothing to do. The board in Y is exceptionally active and vigorous, but that is only because it has a function—management of a municipal home hospital—on which the federal and state governments have not encroached.

Members of the welfare board are not the only local officials who might be supposed to influence the assistance program. Mayors, managers, and councillors process assistance budgets, and they often receive complaints and requests which impel them to inquire into particular cases.

Assistance budgets invite the attention of politicians because of their size but resist cutting because of their invulnerability to local influence. In the 29 cities in 1962, public assistance accounted, on the average, for 14 per cent of municipal expenditures; in 4 cities they were more than a fifth of the budget. The local government actually pays for only a fifth of these costs, but the burden on the local tax rate is still considerable, an average in the 29 cities of $5.64 per thousand dollars of assessed valuation. Even so, the local expenditure is not subject to local choice—at least, not to any choices that politicians are able to discern. In appropriating assistance funds, the city government is meeting an obligation the terms of which are set elsewhere. Neither in X nor Y does the council scrutinize the welfare budget carefully.

When politicians receive complaints from recipients, they take them up with the welfare department, invariably to find that it is "just following the law."

Even if they enter office with the expectation of exercising influence over assistance, welfare board members and politicians soon encounter the obstacle of state rules and cease trying. What they cannot influence, most prefer to ignore. One striking exception is the energetic young mayor of New Bedford, who has taken a keen interest in public assistance—for the purpose of ridding his government of it. In his inaugural address in 1964 he said:

> . . . the relationship between the Commonwealth of Massachusetts and the cities and towns in the welfare field has reached a point where responsibility and control have been separated. The State has taken over control of welfare programs. . . . I have already initiated in the General Court, legislation to the effect that the Commonwealth assume not only the control of the program but also the responsibility for it. . . .

Despite the latent resentment among politicians and presumably among their constituents of exploiters and immoral mothers, neither in X nor in Y have politicians seized upon this seemingly attractive issue. One suggested that the issue was too risky—that no one wanted to attack a program that is admirable in principle and to some extent in practice, and that a distinction between the good, popular and the bad, unpopular features of it might be difficult to sustain in the simplified rhetoric of campaigns. But the usual explanation was the one offered by the mayor of Y: "There is no sense getting involved in something you can't do anything about."

If the differences between X and Y cannot be traced to differences in politically expressed community values, neither can they be traced to differing attitudes among workers. In both cities, workers generally expressed liberal, lenient attitudes. They strongly supported the federally inspired trend toward service-giving, typically defined a good worker as one who is "compassionate," "likes people," is able "not to judge others by themselves," favored more generous grants, and doubted that the assistance program creates dependency, although the answers to this question were usually qualified. Workers generally tend to be more lenient than agency executives; this was especially true in Y, but even in X one worker strongly criticized the agency for failing to do more for clients.

The differences are traceable rather to differences in executive direction—in the kinds of goals defined by executives and in the methods used to achieve them. Agency Y is rule-oriented and internally authoritarian; rigid adherence to rules is the principal demand placed upon workers, and they are subject to constant, close, aggressive supervision. They are expected to obtain executive approval for any action that is not wholly routine, even if it does not entail the expenditure of money; routine actions must be recorded and are scrutinized with care by the agency supervisor. Workers are frequently chastised for departures from routine or failure to perform routine properly. Agency X is client-oriented and internally permissive; within limits imposed by state rules and informal agency practice, workers are free to exercise discretion and to pursue the goal of "helping people." Executives do not intervene aggressively in their activity or chastise them openly. Workers must seek executive approval for services to clients that involve expenditure of funds, but they do this with a high expectation that it will be given.

Executives in Agency Y do not expressly discourage client-serving activity. On the contrary, workers are frequently exhorted to "give services," to "do something" about clients' problems, to include in case records a "treatment plan," which is one of the formal requirements imposed by the state in its effort to encourage a therapeutic, social-work approach to public assistance. The insistence on a treatment plan is characteristic of Agency Y's punctilious observance of rules. Close supervision of workers in Y is intended above all to minimize their exercise of discretion, and it is characteristic of the director of Agency Y not only to deny discretion to others but to deny it to himself. In Agency Y, the practice is to execute state laws and rules that are expressly compelling and to ignore or interpret narrowly provisions that are merely permissive. The effects of this strict, rigidly supervised observance of rules is to discourage client-serving behavior, for a therapeutic, social-work approach to the administration of public assistance depends heavily on the possession of initiative and discretion by the worker who is responsible for the case.

The fundamental difference between X and Y, then, is not that one is lenient toward clients and the other is harsh, or, as professional social workers would say, "punitive." It is that in Y, rules are treated as ends in themselves and interpreted in such a way as to limit the exercise of discretion, whereas in X the client is the principal object of action and rules are frequently interpreted in such a way as to permit discretionary expenditure on the client's behalf. While Agency X is lenient, Agency Y is "legalistic." The effect of these differences in orientation toward rules, and of associated differences in supervision of workers, is, in X, a comparatively generous program, in Y, a comparatively penurious one.

These outcomes are not, however, merely accidental—the unintended consequences of the orientation of particular executives toward rules *per se* and toward organizational subordinates. Agency X is deliberately generous. "We believe," the director said, "in granting any request that is reasonable." Director X is strongly committed to the concept of services, chose to attend the state welfare department's orientation sessions in Boston a few years ago when service-giving began to be stressed, and remains sensitive to the opinions and approval of state supervisory officials. Director Y contrasts sharply with Director X. He is a strong believer in "home rule," does not place a high value on having approval or advice from state officials, and, while not overtly opposed to the recent stress on service-giving, has taken the position that this is something the local agency has always done and done well. Although he does not profess either generosity or penury ("We follow the state standard"), it is unlikely that the exercise of discretion would be so strongly discouraged in Agency Y if the substantive result were not intended. In both cities, the outcomes appear to be those sought by the director, who has a high degree of autonomy within the local government.

Federal and state domination of public assistance policy-making has not altogether eliminated discretion at the local level, but it has shifted the locus of such power as remains there. To the extent that discretion is available locally, it belongs to professional administrators, most of all to the one at the head of the agency. It is his values and preferences that are expressed in the local government's application of state laws and rules.

The impenetrability of state rules is the director's main protection against the influence of others. These rules are not readily accessible, and they are so numerous, so complicated, and change so often that only the most zealous and persistent outsider could hope to master them. Few welfare board members and still fewer politicians make the attempt. The local administrator becomes an unchallengeable authority on what the rules require, not because his mastery of them is perfect, but because it far exceeds that of anyone outside the agency. Whether faced with a general proposition or a particular question, he can deny with a high degree of truth and still higher degree of plausibility that he possesses any choice. Those instances in which he does possess choice are very difficult for another party—be it an AFDC mother or the city's mayor—to detect. The area of local discretion is elusive and always diminishing.

Another source of the director's autonomy is the civil service system. Under federal and state laws, it must cover directors as well as other local administrative personnel. Only when a vacancy occurs—which is rarely, for the tenure of public assistance employees is long in Massachusetts[8]—can a mayor expect to choose his own welare director. Only at that point, initial selection, is the civil service system manipulable. It thoroughly protects incumbents, who are accustomed to watching mayors and managers come and go. The usual retirement age is 70, but there are exceptions even to this. The director in Y is among a small group that is not legally obliged to retire at all. Because he is a World War I veteran and because he elected not to participate in a pension plan, he will hold office until he quits or dies.

Of course, no director is free from local influences, if only because he is himself a product of the community and his own values are to some degree a product of his experience there. Almost without exception, Massachusetts welfare directors are "locals" who have spent all their lives in or near the place where they work and have developed a strong identification with it. They have not moved from city to city in search of career opportunities but have moved through the ranks in one place. Accumulated knowledge, intuition, and experience tell them what other local officials prefer or will tolerate. Agency X might be still more generous if its director did not sense that generosity, if it became public knowledge, would in all probability provoke a reaction. Agency Y's conduct, though considered harsh by some local critics, is approved by the leading politicians there. Y's mayor thought that the director "does a good job of balancing the interest of the taxpayers and the needs of the underprivileged." It would be surprising if the director did not know of this approval and take satisfaction in it. Local preferences, then, must be taken into account, but the expression of them is so muted, infrequent, and disorganized that a direc-

---

[8] See Welfare Administration, U.S. Department of Health, Education, and Welfare, *Public Social Welfare Personnel, 1960*, p. 79. The median length of service of public assistance employees in Massachusetts in 1960 was 13.5 years, highest in the country.

tor is quite free to make his own assumptions about them or even to follow his own preferences on the assumption that they are the community's too—or would be if its knowledge of the rules and the clients were equal to his.

If the director's preferences and values are to be reflected in the conduct of the agency as a whole, he needs not only to be free from outside interference but to control the actions of subordinates. Within the agency, he has several advantages in addition to possession of formal authority. Because administration in the state as a whole is decentralized, within any given (small, local) unit, it is highly centralized. In the smallest towns, it is perfectly centralized. The agency consists of only one person who performs the functions of director, worker, and clerk. In the cities there are several levels of authority. Supervisors at least, and possibly a deputy director and principal or head social work supervisors, intervene between the director and workers. Everywhere but in the biggest city, however, the staff is concentrated in a single building: only Boston has district offices. In most places the agency is extremely short of space. The director works in close proximity to workers and supervisors. In short, he has few subordinates to supervise and supervision of them is relatively easy. Workers are often out of the office, but it is not necessary for him to supervise them directly. Most workers rely heavily on supervisors for direction and advice. For the director, supervising the supervisors is sufficient Agency Y illustrates the extent to which centralization of authority goes even in a medium-sized city. It has only two executives—in addition to the director, a supervisor whom the director chose—and they share the same office.

Despite the seeming rigidities of the civil service system, directors are often able to exercise discretion in the hiring of workers. If the civil service register is exhausted (and it usually is, because examinations are given only once a year and the number of takers does not satisfy the demand), the director may look for workers in any way he chooses—for example, by advertising in a newspaper or by approaching friends and acquaintances. Appointed provisionally, such workers acquire experience that enhances their performance on civil service examinations.

Until recently, training and indoctrination of workers have taken place informally within the agency. New workers learned the job by imitating fellow workers and asking questions. Thus the definition of the job that prevailed within the agency was imparted to the newcomer. This process has been facilitated by the fact that workers function in very close proximity to one another. In the typical agency, they share space in the same room with not more than a few feet separating desks. In such circumstances, uniform practice within the agency is encouraged.

Since 1963, local agencies have lost their monopoly of training and indoctrination. All new workers must undergo four weeks of orientation conducted by the state. In attending, they are not only taught a social-work approach, but they meet workers from other agencies with whom they can compare local practices. This experience gives them a perspective from which to appraise their own agency. In the case of workers in Y, it seems to have left a lasting impression and helps to account for the fact that their attitudes differ from those of the director.

Not all agency heads dominate their agencies. For a variety of reasons, usually having to do with age, health, or temperament, some do not try. In such cases discretion rests with supervisors and workers. In some agencies, workers are able to obtain approval of their proposals and in effect to exercise discretion themselves by selecting the executive from whom approval will be sought. The bigger the agency the more executives there are, and the worker may decide to rely for advice on one who will give the kind of advice the worker wants to hear, even if this means going outside channels.

Agencies can be divided into three categories, according to the degree of authority exercised by the director and the character of agency conduct: centralized-restrictive; centralized-lenient; and decentralized, in which the differing values of individual supervisors and workers find expression. Decentralized agencies are likely to fall near the middle on the leniency scale, though there is a tendency for decentralization to produce leniency, because workers (not being responsible for defense of the agency's budget and being immediately subject to the clients' requests) are less concerned about costs than directors. The smallest agencies are divided between the first two categories and the largest are divided among all three, depending on the values of the director and his will and capacity to impose them on the agency as a whole.

THE EFFECTS OF COMMUNITY TYPE

Cases X and Y—and other cities observed but not reported on here—indicate that the agency head plays the crucial role in determining the character of local actions. Intercity differences that depend upon the use of local discretion appear to derive from differences in the values and administrative practices of agency chief executives.

However, cases X and Y do not show that socioeconomic variables have no effect on administrative outcomes. Outcomes in the two cities being demonstrably different, they were selected for study in order to answer the question, why are some places more lenient than others? The fact that they are quite similar socioeconomically made it inevitable that the answer would be found elsewhere than in socioeconomic variables. To test the effect of such variables, it would be necessary to survey cities of contrasting socioeconomic type. This has not been done, but preliminary observation suggests that the results would be negative.

An attempt has been made, by aggregating data for certain groups of cities and towns, to test the effect of urban-rural and class differences. Although average measures of output for any individual place with a small caseload cannot be treated as even partially valid indicators of the degree of leniency, it is possible to group the smaller places and measure the tendency of the group.

The remaining 322 cities and towns (351 less the 29 with AFDC caseloads of at least 100) were divided into those with populations above and below 10,000. These groups will be referred to as "urban towns" and "rural towns." Those above 10,000 were divided into three subcategories: those with median family incomes of $7,000 or more ("upper-middle-class"); those between $6,000 and $6,999 ("middle-class"); and those of $5,999 or less ("lower-middle-class"). The number of towns and the number of cases as of July 1965 in each group are shown in Table 3.

From aggregate data for all four groups, and for the 29 cities in the original group, averages were calculated for total payment, cash grant, medical vendor payment, and nonmedical vendor payment per recipient per month. (The medical vendor payment is uncorrected for intercity differences in hospital rates.) The results, summarized below in Table 4, show no consistent correlation between class and the expenditure measures, and although the rural towns rank lowest on all measures, the differences between them and the urban towns are not large. They are not so large as to preclude the possibility that they could be accounted for by differences in caseload characteristics, other variables unrelated to agency behavior, or flaws in the data. The absence of a significant expenditure difference is surprising, for the smallest towns have a reputation for penury. The reason alleged for this is that their programs are dominated by the selectmen (the general governing body of the town)

TABLE 3

	Rural	Urban UMC	Urban MC	Urban LMC
Towns	232	34	33	23
AFDC cases	1,797	840	1,251	1,079
Cases per town	7.7	24.7	37.9	46.9

TABLE 4

	Total Payment per Recipient per Month[a]	Cash Grant per Recipient per Month	Medical Vendor per Recipient per Month	Nonmedical Vendor per Recipient per Month
Rural Towns	$45.88	$40.16	$6.28	$0.29
Urban LMC	47.07	40.93	6.79	0.36
Urban MC	47.05	40.53	7.18	0.29
Urban UMC	47.95	41.56	6.87	0.31
29 Cities	49.95	43.99	5.79	0.69

[a] The three component expenditures—cash grant, medical vendor, and non-medical vendor—add up to more than the total expenditure because two different figures for the recipient population have been used in computing the averages. In computing the cash grant and nonmedical vendor averages, recipients who received assistance only for medical care were excluded. These medically indigent persons, poor enough to qualify for medical assistance but not so poor as to receive AFDC regularly, accounted for 1.4 per cent of all recipients in 1965.

which often functions also as the board of public welfare. The selectmen, in other words, wear two hats, and they are widely suspected of letting their interest in conserving the taxpayers' money take precedence over care of the poor. The figures do not bear this suspicion out, and my own impression—based on a very few interviews with small-town directors—is that selectmen are often just as baffled by public assistance rules and just as uninterested in them as big-city mayors. Even in the small town, the director is likely to have considerable freedom, though the potential for intervention by elected officials is unquestionably greater in the small places than the large ones.

The most striking fact revealed by the table is that the cities with the largest caseloads, which are nearly the equivalent of the largest cities, rank above all other places by a fairly wide margin.[9] They spend twice as much as other places on nonmedical vendor payments and $2.50 to $3.00 more on the cash grant. But the table exaggerates the extent to which these differences are systematic; for the city averages are biased by the liberality of the biggest single city, Boston, which has 35 per cent of the AFDC cases in the state and 43 per cent of those in the 29 cities. When the city data are aggregated, the resulting averages are much affected

[9] Except for the anomaly of vendor medical payments, which can probably be explained largely by the fact that these cities rely heavily on municipal hospitals, whose rates are lower than those of private hospitals.

by the Boston cases. The extent to which this occurs can be seen by comparing the averages derived from the aggregated data (a method that gives Boston proportionate weight) with the averages derived from city averages (a method that gives Boston the same weight as the other 28 cities). The cash grant average jumps from $42.86 to $43.99, and the nonmedical vendor average from $0.35 to $0.69. The aggregate city average for nonmedical vendor payments exceeds the amount spent by 26 of the 29! Thirteen of the cities spent no more on nonmedical vendor payments than the rural-town average.

Boston is unusually lenient in the granting of furniture, appliances, and moving expenses. It spent over $40,000 a month on such items for AFDC recipients in 1965. It has a rule that no client shall go without utilities overnight and to prevent disconnections spent an estimated $50,000 in 1965 on AFDC clients' overdue bills. It frequently adds to the cash grant with allowances for errand service, transportation, laundry expenses, and other special needs.

Boston's leniency is related in many ways to its large size. Much authority to grant special needs must be decentralized, and the more decentralized it is the more freely it is used. The department is more subject than others to professional influence from within (some of the top executives have degrees in social work) and to pressure from liberal reformers without, including civil rights groups, professional social workers, militant collegians, AFDC mothers whom the social workers and collegians have organized, and a daily press that is always eager to expose the errors of central-city government to suburban readers. Other cities, only a fraction of Boston's size and less richly endowed with universities, social-service organizations, and other civic institutions, barely feel the effects of such activity.

Size, then—or, more precisely, certain factors associated with it—may produce leniency, but the data suggest that the correlation, if any, between size and leniency is not a simple one. It is not true that the larger the community, the more lenient its welfare administration is likely to be, but the case of Boston suggests that there may be a "threshold point" beyond which the effects of size make themselves felt.

If Boston were removed from the table and treated as a unique, deviant case, the apparent differences between cities and other types of communities would be much reduced, but even if this change were made, the grouped data would not show that uniformity exists. Each of the groups has its "X's" and its "Y's," individual deviants from the group norm. What the data do show is that differences are not clearly and systematically related to community type.

It may be that such a relation has never existed. That differences in type of community should cause differences in welfare policy outcomes may be a social scientist's fantasy. Or it may be that such a relation would exist if federal and state attempts to impose uniformity had not been highly successful. In this view, the relation does exist "naturally" but has been artificially suppressed by influences external to the local community.

One way to test these explanations and to appraise the impact of federal and state intervention is with data from that segment of the public assistance program, general relief, that remains under local control. The results (see Table 5) suggest that the second explanation is more nearly correct.

TABLE 5

	General Relief Grant per Recipient per Month
Rural Towns	$21.01
Urban LMC	24.79
Urban MC	25.79
Urban UMC	36.97
29 Cities	38.75

There are marked urban-rural differences in general relief grants per recipient, and the size of grants is positively correlated with class.

CONCLUSION

Federal and state intervention has not produced uniformity among local agencies in the administration of AFDC in Massachusetts. There are differences in local administrative outputs; and, to some extent, these are not the result of objective differences in the circumstances of clients but result from differing degrees of leniency among local administrators, some of whom are more disposed than others to use such discretion as they have in ways that increase benefits to clients.

But if the federal ideal of uniformity is less than perfectly realized, federal and state intervention has nonetheless had a profound impact on local assistance administration. The local agencies' scope for the exercise of discretion is small, and differences in outputs are much less than they would be if local values were freely expressed through the processes of local politics. They are but slightly, if at all, related to community type. The existing differences are marginal; they are no greater than what can be expected when a single set of rules, highly complex yet nowhere near complex enough to cover all contingencies, is administered by many individuals of differing personalities, beliefs, and administrative capabilities.

If the federal ideal were realized, rules would be applied everywhere in the same way. Like cases would be treated alike no matter where they lived. One way of achieving this goal is by the elaboration of rules, which, as they increase in number and detail, reduce the discretion of administrators. This is the method that the federal and state governments have used, and it has fundamentally changed the public assistance program in Massachusetts. Thirty years ago, the distribution of assistance was determined by local politics—mainly by politics in the ideal sense, as the process by which community values are expressed and allocated, but also by politics in the vulgar sense, as the allocation of favors on the basis of friendship and self-interest. Today, assistance is in general distributed in accordance with rules that originate elsewhere than in the local community and are little affected in content or application by local politics in any sense.

On the other hand, it is impossible to write rules of such detail that they cover all possible contingencies with respect to all cases. Nor is it desirable, according to the doctrines of professional social work, which guide welfare administrators at the state and federal levels. The individual worker must be free to adapt "treatment" to the particular case, according to the principles of "good casework."

Leaving discretion in the hands of caseworkers and other local administrators need not compromise the principle of uniformity if they share a common set of values, derived from a shared professional code or ethic. Like cases would still be treated alike, according to the prescriptions of the code. A second method of achieving uniformity is the propagation of such a code among public assistance personnel at the local level.

The federal and state governments are using this method too. The federal requirement, imposed in 1940, that employees be covered by a merit system was a first step. In 1964 the federal Welfare Administration issued a directive requiring that henceforth all public assistance workers and supervisors have college degrees. The welfare commissioner has stated that graduate training should be a prerequisite for advancement above the rank of worker. Presumably, the federal ideal is a public assistance bureaucracy in which every member has a master's degree in social work. In the meantime, the content of "professionalism" is disseminated through literature and in-service training programs.

Professional values have to some extent been imparted to local activity by the content of state-prescribed rules, but extension of them will be complete only when local administrators have themselves undergone formal indoctrination and learned to apply professional values in the matters that remain discretionary. This time is a long way off. A nationwide survey in 1960 showed that only 3.3 per cent of all state and local public assistance personnel and less than .05 per cent of caseworkers had master's degrees in social work. In Massachusetts only 13.9 per cent of public assistance employees had graduated from college, and 42.7 per cent had not attended college.

In brief, "politics" has disappeared from the public assistance administration in Massachusetts, with much prodding from federal and state administrators, but "professionalism," despite their efforts, has not yet taken its place. We are in a transitional stage where the personal values of administrators are of great importance.

J. DAVID GREENSTONE

PAUL E. PETERSON

Reformers, Machines, and the War on Poverty

Scholars of American politics have been notably solicitous about the interests of the urban lower classes. In particular, this concern has focused on the class bias of machine and reform politics. The resulting analyses, however, have produced conflicting interpretations. While political scientists of an earlier generation believed political reforms weakening urban machines would benefit the poor, some contemporary scholars have taken an opposite view. Nevertheless, these two perspectives share certain common insights which we propose to apply in examining four local community action programs of the Economic Opportunity Act of 1964. Specifically, we seek to explain the variation in the initial implementation of the poverty program among the nation's four largest cities—New York, Chicago, Los Angeles, and Philadelphia.[1] We shall argue that in its first two years (1964 to 1966) the war on poverty pursued two goals which were not entirely compatible: (1) to end economic poverty by distributing various material perquisites, and (2) to end the virtual exclusion of low-income groups from political life by distributing power. Consistent with their historical backgrounds, reform cities were generally more successful at distributing power, while machine cities were more adept at distributing material perquisites.

[1] The authors wish to express their appreciation to the Russell Sage Foundation for making possible this research. Mr. Peterson would like to express similar appreciation to the Woodrow Wilson Foundation.

267

MACHINES AND REFORMERS IN CITY POLITICS

S. M. Lipset has observed that:

. . . the political intellectual, the man of ideas, is nowhere very interested in defending inconsistencies, and every *status quo* is full of inconsistencies. Only by attacking the limitations of his political and social order can he feel he is playing a fruitful creative role.[2]

Accordingly, scholarly analyses of city politics have varied inversely with the changes in the political life of American cities. From the late nineteenth century to World War II, political scientists viewed with alarm the concentration of power achieved by organized political parties.

While relying on the votes of the poor, the machine cooperated in their exploitation by more privileged groups. As Ostrogorski said, machines only insured "the power of plutocracy . . . in the political sphere."[3] Machines often provided large personal profits to individual racketeers, speculators, and businessmen. Basing their power on control of lower-class voters, the most successful members of the machine emulated their robber baron contemporaries in accumulating personal fortunes. The motto of George Washington Plunkitt of Tammany Hall, a master at procuring "honest graft," was "I seen my opportunities and I took 'em."[4] Merton, himself a sympathetic interpreter of the machine, observed that one of the machines' "latent functions" was to provide a locus of political power with which the business community could negotiate.[5] At the height of their power, moreover, machines were not noted for instituting massive new government services that redistributed wealth collectively to the urban lower classes.

Throughout the twentieth century, however, the machine's power has declined, partly in response to rising education and income levels, the greater political influence of the news media, and increasing bureaucratization of welfare programs. Of equal importance, the reform movement, adopting the views of Ostrogorski and others, weakened or destroyed

[2] S. M. Lipset, *Political Man* (Garden City, New York: Doubleday, 1963), p. 345.
[3] M. Ostrogorski, *Democracy and the Organization of Political Parties,* II (Garden City, New York: Doubleday, 1964), p. 299. Ostrogorski was far from alone in his bitter attack on the "machine." Other critics include James Bryce, *The American Commonwealth II* (New York: MacMillan, 1895), and Roy V. Peel, *The Political Clubs of New York City* (New York: G. P. Putnam's Son, 1935). Of great influence in this generally critical attitude toward the party organization was Robert Michels' *Political Parties* (New York: Collier Books, 1962), which documented the oligarchical tendencies within the Social Democratic parties in Europe.
[4] William L. Riordan, *Plunkitt of Tamanny Hall* (New York: E. P. Dutton, 1963), p. 3.
[5] Robert Merton, *Social Theory and Social Structure* (Glencoe, Illinois: Free Press, 1957), pp. 75–76.

party machines by transforming the politics of many American cities. Reform innovations, such as nonpartisan elections and a civil service merit system, often eliminated patronage and reduced graft, two bulwarks of machine power.

As the power of the machine declined, new scholars, in keeping with Lipset's observation, began to depict favorably the political machine as the articulator of the wants of the immigrant, working-class population. They have cited the views of machine politicans such as Plunkitt himself:

> If a family is burned out I don't ask whether they are Republicans or Democrats, and I don't refer them to the Charity Organization Society which would investigate their case in a month or two and decide they were worthy of help about the time they are dead from starvation. I just get quarters for them, buy clothes for them if their clothes were burned up, and fix them up till they get things runnin' again. . . . The consequence is that the poor look up to George W. Plunkitt as a father, come to him in trouble and don't forget him on election day.[6]

Simultaneously these contemporary scholars saw reform innovations favoring more privileged groups in the community, and they began to view this movement with increasing suspicion.[7] Hofstadter's influential analysis of *The Age of Reform* spelled out the middle-class, Protestant basis of the Progressive movement as a reaction to the value system of lower-class, Catholic immigrants. Hofstadter argued that:

> On one side [the reformers] feared the power of the plutocracy, on the other the poverty and restlessness of the masses. But if political leadership could be firmly restored to the responsible middle classes who were neither ultra-reactionary nor, in T. R.'s phrase, "wild radicals," both of these problems could be met.[8]

Political scientists have added that the reformers' structural changes contributed to middle-class domination of city politics. In nonpartisan elections the lower-class voters, lacking a middle-class sense of civic duty, stayed at home, for there was no organization to stir them to political action.[9] Business-oriented civic associations and daily newspapers became the bulwark for many office-seekers.[10] City-wide elections favored politi-

---

[6] Riordan, pp. 27–28.

[7] For example, see Eugene C. Lee, *The Politics of Nonpartisanship* (Berkeley: University of California Press, 1960).

[8] Richard Hofstadter, *The Age of Reform* (New York: Alfred A. Knopf, 1959), p. 238.

[9] Lee, pp. 139–140. Also, Oliver P. Williams and Charles R. Adrian, "The Insulation of Local Politics under the Nonpartisan Ballot," *American Political Science Review*, LIII (December, 1959), pp. 1059–1061.

[10] Edward C. Banfield and James Q. Wilson, *City Politics* (Cambridge: Harvard University Press, 1963), Chap. 21.

cians with independent sources of income.[11] The new city manager, with an invariably middle-class background, found himself most at home with leaders in the business community.[12] Finally, these new scholars have suggested, the declining strength of party organizations encouraged an interest group politics, often favorable to large corporations and businessmen's associations.

In our view this critique of reform movement overlooks several of its important attributes. Certainly the reformers initially attacked both the machine's accumulation of power and its defense of privilege. Social workers such as Jane Addams and muckrakers such as Lincoln Steffens called for public policies which would improve the position of the entire lower class.[13] Earning the enmity of businessmen, they advocated better schools, garbage collection, and other services in poor areas, as well as the abolition of sweatshops and child labor. The comparatively progressive policies of Theodore Roosevelt and Woodrow Wilson were in many ways the product of the reform movement.

Nevertheless, the reformers' paternalistic attitude toward the lower classes, their reluctance at times to engage in ethnic politics, and their indifference to the immediate material needs of individual immigrants prevented them from developing a reliable lower-class constituency. Following World War I, the social conscience of the reform movement was conspicuously absent; as the entire political system became more conservative, reform focused more on corruption than on general social ills. Reformers were more successful in altering political structures than in redistributing wealth to the poor.

We conclude that each scholarly tradition was more accurate in its critique than in its defense, since both the machine and the reform movement had conservative consequences. For businessmen "on the make," machine politics provided franchises and special privileges. For their better established successors good government seemed both efficient and morally praiseworthy.[14] The machine *controlled* the lower-class vote, while somewhat later the reformers' structures *reduced* it. By drastically reducing party competition each protected vital business interests from significant political interference. Their consequences were similar to those of the

[11] Lee, pp. 76–84.
[12] See John Bartlow Martin, "The Town That Tried Good Government," in Edward C. Banfield, ed., *Urban Government* (New York: The Free Press of Glencoe, 1961), pp. 276–284. Reprinted from *Saturday Evening Post,* October 1, 1955.
[13] The best source for the views of the muckrackers is in Lincoln Steffens, *The Autobiography of Lincoln Steffens* (New York: Harcourt, Brace & World, 1931).
[14] Banfield and Wilson, p. 265. For an illuminating discussion of the transformation of American reform, see Michael Rogin, *The Intellectuals and McCarthy: The Radical Spectre* (Cambridge, Mass.: MIT Press, 1967), Chap. 7, especially pp. 202–207.

one-party system of 1896, which, as Burnham shows, dramatically reduced and disoriented the electorate in national and state party politics from 1900 to the New Deal.[15]

Indeed, once the question of favoritism for particular classes is set aside, the two scholarly traditions actually agreed on certain differences between machine and reform politics in the allocation of political values. Machines directly tied their governmental outputs to the maintenance of their organization. The machine was willing and able to distribute governmental resources to individuals, but the criterion of distribution was not whether the individual was deserving or fell within the category prescribed by the relevant law but whether the individual contributed to the political success of the organization. For rich and poor alike, partisan political criteria were the universal standards for distributing values. At the same time, the patronage system meant that party campaign workers staffed the city's administrative bureaucracy. Both policy and personnel practices, then, served to obliterate the distinction between input and output structures. The early scholarly tradition condemned these practices of the machine as an abuse of governmental power for partisan aggrandizement. Modern revisionists have noticed the importance of these practices in serving the individual wants of an immigrant, working-class population.

The reformers, on the other hand, have studiously tried to separate the input and output structures. By eliminating patronage, exposing graft and corruption, and rigorously adhering to the letter of the law, reformers have consciously conformed to explicit legal criteria in distributing government outputs.[16] These steps, Ostrogorski believed, would mean the reinvigoration of democratic processes. More recent scholars, however, have emphasized the advantages such changes confer on the middle and upper classes in urban politics.

These differences in style had important consequences for urban political systems as such. Whereas the machine centralized political power in its own hands, the reformers overtly attempted to disperse this power to "better government" civic associations, the civil service, the press, and (through the initiative, referendum, and recall) to individual voters. Many reformers did centralize formal *authority* in the hands of the mayor or city manager, but these steps by no means fully offset the overall dispersion of power.[17] In sum, the machine concentrated power among a few of its own leaders, while the reform movement dispersed power to a wide range of individuals, groups, and agencies.

[15] Walter Dean Burnham, "The Changing Shape of the American Political Universe," *American Political Science Review,* LIX (March, 1965), pp. 7–28.
[16] An excellent discussion of the goals of these municipal reformers can be found in Banfield and Wilson, pp. 138–186.
[17] Banfield and Wilson, pp. 101–111.

But this very dispersion of power produced latent consequences unforeseen by reformers. The centralization of political power by the machine enabled city officials to move quickly and decisively in obtaining and distributing material resources. The machine's distribution system may not have followed Weberian bureaucratic norms, but its cohesive, centralized structure of power enabled it to move efficiently in overcoming political barriers to the establishment of governmental programs. Once a program was established, a reform administration could employ a bureaucracy committed to principles of scientific and efficient management. But these more pluralistic reform regimes found it exceedingly difficult to secure agreement among all the centers of power who could veto their suggestions. The dispersal of power, in other words, made the rapid distribution of resources more difficult.

Both the upper-class bias common to reformers and machines and the scholarly consensus on the different political processes characteristic of machine and reform politics suggest that the real dispute between the two scholarly perspectives reflects an underlying clash of political values. Defenders of reform voiced a nineteenth century Jeffersonian optimism: widespread political participation meant good citizenship, vigorous republican government, and sound public policies.[18] The vote buying, vote stealing, and sheer organizational strength of the machine led Ostrogoski to argue that:

> Where the Machine is supreme, republican institutions are in truth but an idle form, a plaything wherewith to beguile children. . . . It is no longer "a government of the people, by the people, and for the people."[19]

Reflecting the pessimistic theories of mass society characteristic of the twentieth century, critics of reform believed the public interest is better served by limiting direct mass participation in policy making. This theory argues that such direct participation as that currently generated by interracial tension too often leads to confrontations that not only make compromise of diverse interests difficult but also prevent political leaders from disregarding "public opinion at crucial moments when public opinion, or intensely moved parts of it, is out of line with long-term national interests."[20] By contrast, machine control of the electorate through ethnically balanced tickets and material payoffs to individual voters reduces

---

[18] An excellent statement of the views of nineteenth-century theorists on democracy as distinguished from the pluralist viewpoint is given in Jack L. Walker, "A Critique of the Elitist Theory of Democracy," *American Political Science Review*, LX (June, 1966), pp. 285–296.

[19] Ostrogorski, p. 300.

[20] Banfield and Wilson, p. 345.

the accessibility of city officials to mass pressures and any temptation to demagoguery.[21]

We contend that these three considerations—the upper-class bias of both reform and machine politics, the different approaches of machines and reform to distributing power and perquisites, and the scholarly disagreement on the proper political role of the mass population—are all relevant for analyzing the intercity variation in the administration of community action programs.

COMMUNITY ACTION PROGRAMS: DISTRIBUTING MATERIAL PERQUISITES OR POLITICAL POWER?

The war on poverty is one of the clearest examples of a post-New Deal welfare state program designed to redistribute material perquisites to lower-class citizens. Community Action Programs were expected by law to give "promise of progress toward eliminating poverty . . . through developing employment opportunities, improving human performance, motivation, and productivity, or bettering the conditions under which people live, learn, and work."[22] But the legislation conceived of poverty as a political as well as an economic condition. In a celebrated phrase it required that Community Action Programs be "developed, conducted and administered with the maximum feasible participation of residents of the areas and members of the groups served."[23] While some personnel in the federal Office of Economic Opportunity (OEO) were concerned only with finding more jobs and better services for the poor, other officials, together with certain civil rights organizations, saw this "maximum feasible participation" phrase as an opportunity to increase the political power of the Negroes and other disadvantaged citizens. According to the OEO's Community Action Workbook, for example, a "promising method" of implementing "maximum feasible participation" was "to assist the poor in developing autonomous and self-managed organizations which are competent to exert political influence on behalf of their own self-interest."[24]

According to the view that poverty was a political as well as an eco-

[21] Edward C. Banfield, *Political Influence* (New York: The Free Press of Glencoe, 1961), pp. 260–262.

[22] U.S. Congress, *An Act to Mobilize the Human and Financial Resources of the Nation to Combat Poverty in the United States,* Public Law 88-452, 88th Cong., 2nd Sess., 1964, p. 9.

[23] *Ibid.*

[24] Office of Economic Opportunity, *Community Action Workbook* (Washington, D.C., 1965), III. A. 7.

nomic condition, low-income groups lacked financial resources, social prestige, and easy access to decision makers. Such groups continue to be known for low voter turnout where parties are weak and for the ease with which their vote can be "controlled" by strong party organizations. Most important of all, the poor have had few autonomous organizations which articulate their collective demands and maximize their electoral influence—requisites for becoming more than a "potential group" in urban politics.[25] "Maximum feasible participation" was thus interpreted as the organized and active pursuit of political power.

In practice, the OEO was constrained by the decentralized American political system to administer the "war on poverty" through local community action agencies, keeping for itself only the power to choose among the proposals submitted for its approval. Consequently, its attempt to disperse political power to the poor was confined to the formalistic requirement that representatives of the poor—chosen "whenever feasible" in accord with "traditional democratic approaches and techniques"—comprise approximately one-third of the policy-making body for local Community Action Agencies.[26] Although its actual effectiveness is far from clear, presumably this requirement would stimulate the growth of indigenous organizations, increasing the political power of the poor. In any case all the major political actors—the OEO, the big city mayors, reform and civil rights leaders, interested Republicans, congressmen, and articulate members of neighborhood groups—regarded the question of representation as potentially significant.

The duality of viewpoints within the OEO was reflected in the pattern of cleavage in all four cities. In 1964 and 1965 the four incumbent mayors regarded poverty as an economic condition and resisted those in OEO who sought to disperse political power. Initially, the mayors each formed a committee to centralize decision-making among key members of their administrations.[27] When OEO sought to disperse political power by including representatives of the poor on the policy-making body, Wagner of New York articulated the common mayoral reaction in testimony before a congressional committee.

When I testified a year ago, I urged that the local governing bodies, through their chief executives or otherwise, should have the ultimate authority, as

---

[25] This term is drawn from David Truman, *The Governmental Process* (New York: Alfred A. Knopf, 1965), pp. 511–516.
[26] Office of Economic Opportunity, *Community Action Program Guide* (Washington, D.C., 1965), I, p. 18.
[27] Characteristically, the extreme decentralization in Los Angeles required the participation of other governmental units even at the beginning.

they have the ultimate responsibility, for . . . the conduct and operation of the anti-poverty program.[28]

Similarly, the executive committee of the U.S. Conference of Mayors resolved:

> *Whereas,* no responsible mayor can accept the implications in the Office for Economic Opportunity Workbook that the goals of this program can only be achieved by creating tensions between the poor and existing agencies and by fostering class struggle; . . . NOW THEREFORE BE IT RESOLVED that the Administration be urged to assure that any policy . . . assure the continuing control of local expenditures relative to this program by the fiscally responsible local officials.

The mayors had sound political reasons for viewing poverty as an economic condition. An antiquated tax structure had created severe financial burdens for city administrations. Demands of the poor and minority groups for costly changes in education, for expansion of hospitals and public health services, expensive vest pocket parks, and improved welfare services obviously strained local budgets.[29] Mayors, therefore, welcomed federal aid designed to alleviate the economic plight of slum residents who might resort to direct action and even violence. To be sure, the Community Action Program was far from a complete answer to these problems; the resources available were both too few and too restricted.[30] As practical politicians, however, the mayors regarded any program alleviating economic poverty as better than none; besides, the accompanying publicity would give at least the appearance of action.

By contrast, attacking poverty as a political condition appeared exceedingly risky. Dispersing power to the poor would only increase their demands for more governmental services, which were already in short supply. Even worse, these demands of new autonomous organizations would inevitably antagonize other urban interests and weaken the mayor's own position. Complaints about police brutality and calls for a civilian review board had aroused the ire of police departments. Public housing authorities had often objected to the formation of militant tenant groups demanding

---

[28] U.S. Congress House Committee on Education and Labor, *Hearings, Examination of the War on Poverty Program,* 89th Cong., 1st Sess., 1965, p. 483.

[29] For an interesting discussion of needs, plans, and possibilities in New York City, see Thomas P. F. Hoving, "Think Big About Small Parks," *New York Times Magazine* (April, 1966), pp. 12–13, 68–72. Changes in welfare policies—and their costs—are discussed by Richard A. Cloward and Frances Fox Piven, "The Weight of the Poor: A Strategy to End Poverty," *The Nation,* CCII (May, 1966), pp. 510–517.

[30] The legislation forbade use of poverty funds for primary or secondary education and limited OEO's educational activities largely to Head Start programs.

more responsiveness to the wishes of residents. Principals and teachers generally opposed public pressure against traditional educational practices, especially when generated by poorly educated slum dwellers. Real estate interests bitterly attacked picketing, demonstrations, and rent strikes aimed at better maintenance of tenement housing. Private welfare agencies have disliked the development of autonomous organizations that compete for clients, funds, and staff. Few mayors desired to antagonize any of these entrenched interests; independent militant groups spawned by the war on poverty might well arouse most of them.

Understandably, then, the mayors in all four cities simultaneously conceived of poverty as a purely economic condition and sought to centralize power during the initial development of the poverty program. In each case, however, the chief executive encountered opposition from those who conceived of poverty as a political condition as well. These included neighborhood organizations, certain settlement houses, the more militant civil rights groups, liberal Republicans, and, in Philadelphia, some industrial unions. When OEO revealed that it would withhold funds until the poor were represented on policy-making bodies, these groups held meetings and rallies calling for dispersal of power to the poor. Significantly, their demands were also articulated by two supporters of the reform tradition, who, as our previous analysis would suggest, favored a further dispersal of political power. Newspapers, almost everywhere a stalwart of reform, were particularly important, since one of the few weapons neighborhood groups have is the unfavorable publicity that they can create for the mayor. News stories focusing on the roll of the poor and official reluctance to involve them were published in each of the four cities. The *Philadelphia Bulletin* devoted an average of eight column inches per day to the controversy in the four months before the final decision on the structure of the program. In each of the cities except Los Angeles, at least one major paper backed the critical view of the mayor expressed in news stories with editorial support of varying enthusiasm for participation by the poor.

Favorable coverage of the demands of neighborhood groups reflected the traditional commitment of newspapers to the reform cause.[31] More specifically, in New York City, Republican Congressman John Lindsay had announced as a reform candidate for mayor. Many newspapers were preparing to support him against Democrat Paul Screvane, the man Wagner had chosen both to supervise the poverty program and to succeed him as mayor. In Philadelphia both the major newspapers were in bitter

---

[31] Another motive may have been the penchant of newspapers for controversial issues which boost circulation. But since any number of problems can be made controversial and interesting to newspaper readers, this goal is not a sufficient explanation for newspaper focus on this particular issue.

opposition to the former Democratic ward politician who had become mayor. In Chicago the Republican-oriented newspapers were willing to attack the mayor, although his formidable political strength made them hesitant and selective in their criticism.

Still more vigorous support for representation of the poor was provided by the liberal political clubs who had become the major organizational arm of the reform movement. Not only were the clubs ideologically committed to dispersion of power, but in all four cities they were to some degree antagonistic to the incumbent mayor. In both Los Angeles and New York relatively pro-reform candidates had emerged in the forthcoming mayoral election. The Independent Voters of Illinois and their one alderman on the council vigorously supported participation by the poor in Chicago's program. Philadelphia's more influential Americans for Democratic Action, which disliked Mayor Tate's ward politician image, lent the neighborhood groups office space, staff, materials, and the prestige of a middle-class, intellectual organization. Of the four Los Angeles area congressmen who opposed Mayor Yorty's efforts to assure official control of the program in early 1965, three were identified with the reform clubs in California. (The fourth represented the one Negro district in the state.)

The reformers had their greatest impact in New York City. Democratic Congressman William Fitts Ryan, a leading Manhattan reformer who had recently announced his candidacy for mayor, attacked his rival Paul Screvane for "visions of a new patronage pool." In the spring of 1965 Ryan called for representation of the poor on the policy-making body and for administration of the programs by "democratically" selected local boards. That summer, the Republican candidate, Congressman John Lindsay, adopted the campaign style of a reformer, charging that "City Hall is now setting up a structural monstrosity" and favored instead "prompt steps to increase representation of the poor on the Council."[32]

The dual commitment within the OEO to both economic and political solutions to poverty thus became a matter of open political controversy in the four cities. The mayors held to the conception of poverty as an economic condition and sought to consolidate their power over this program. By contrast, reform groups saw poverty as in part a political condition and supported the dispersal of power to representatives of the poor. Resolution of this conflict varied according to the relative strength of

---

[32] A third Congressman from New York City who contributed to the debate on the poverty program was Adam Clayton Powell. While he cannot be considered part of the reform movement, his political style (which relies on issues rather than on patronage for support) resembles that of the reform clubs. See James Q. Wilson, "Two Negro Politicians: An Interpretation," *Midwest Journal of Political Science,* IV (November, 1960), pp. 349–369.

the combatants. This in turn led to intercity variations in the outputs of the community action program.

INTERCITY VARIATION: THE ATTACK ON POLITICAL POVERTY

In light of the contrasting machine-reform traditions in the four cities, we hypothesized a considerable variation in the local poverty programs funded by OEO. The federal agency would presumably be most successful in distributing power to the poor in reform cities, given the existing dispersion of power. On the other hand, machine cities would probably be more successful at distributing material resources. In order to test this hypothesis, we first assessed the city's *dispersion of political power,* which reflected the existing strength and past success of the reform movement. As an index of power dispersion, we used the strength of the dominant party organization, which in all cases was that of the Democratic party. The four cities were ranked as follows: Chicago (with the strongest organization), Philadelphia, New York, and Los Angeles.

Both qualitative and quantitative data support this rank ordering. In Chicago the mayor was also the leader of the dominant Democratic organization, whose loyal congressional delegation was noted for unrivaled cohesion. Reformers were chronically ineffective, while the Republicans had been an insignificant force in city politics since the 1930's. The Philadelphia Democratic organization also had a relatively united congressional delegation, and in the 1963 mayoral election the organization had elected James H. J. Tate, a ward committeeman, as mayor. On the other hand, the organization lacked Chicago's extensive state and local patronage and after his election Mayor Tate and the party chairman fell into a dispute, reflecting in part the greater influence (compared to Chicago) of reform Democrats in city-wide elections. In 1965, a Republican-reform candidate was elected as the city's district attorney with the support of the Americans for Democratic Action; and in 1966 an independent gubernatorial candidate came within a few thousand votes of carrying Philadelphia in the Democratic primary.

The New York Democratic party organization had so much less cohesion than either Chicago or Philadelphia that the opposition attacked the "bosses" rather than the "boss." Not only had reformers periodically been elected mayor since 1902, but the organization was defeated in the mayoral race both in 1961 by a reform-supported coalition within the party and in 1965 by a Republican-reform opposition. Moreover, the New York Democratic congressional delegation had little unity. The Los Angeles party organization was even more feeble. The city was governed by a nonpartisan mayor who had limited formal powers and operated under a rigid merit system. Nor were state and county governments sources of significant patronage. Legally handicapped in local politics and denied all effective patronage, the formal party organization could not be meaningfully compared even to their New York counterparts. By 1964, the informal Democratic organizations, which had developed over the previous twelve years, were quarreling with each other and were themselves internally divided.

Since these factors have also affected electoral behavior, analysis of election data confirm these considerations. V. O. Key once noted that:

> . . . among the sure districts for each party, there were many in which the organization was so strong that no aspirant dared challenge its man in the primary while in many others the organization was so weak that a primary fight could occur.[33]

Drawing on Key's insight, we examined in each city both the number of candidates per available Democratic nomination to the lower house of the state legislature and the proportion of the vote obtained by the losing candidates. We thus obtained an index of the strength of the organization by considering perhaps the very core of its political power—its ability to monopolize effectively the path to public office.[34] The calculations

[33] V. O. Key, Jr., *American State Politics* (New York: Alfred A. Knopf, 1963), p. 181. Wilson, too, has noted, "The absence of real primary contests is probably as good an indication as any of the power—both actual and imputed—of the machine." James Q. Wilson, *Negro Politics* (New York: Free Press of Glencoe, 1960), p. 45.

[34] The number of candidates aspiring for political office is, to be sure, related to the probability of the party's winning the general election. Where the party has little chance, there may be only one candidate, however weak the organization. But here we are comparing the dominant party organization in each city. Since the Democratic party does well in all four cities in general elections, there are few "safe" Republican seats; thus, this factor should not pose a problem for our intercity comparison.

Special calculations were required in the multimember districts in Philadelphia and Chicago. Here the number of candidates running was calculated in terms of the available Democratic nominations. If three candidates ran in a two-member district, only one race was said to be contested. If four or more ran, both seats were considered contested. In calculating the percentage vote for losing candidates in these two-member districts, the procedures were as follows. Where three candidates ran, the opposition's percentage was calculated by taking the vote of the losing candidate as a percentage of the combined vote of this losing candidate and the weakest winning candidate. Where four or more candidates ran, the percentage for one race was calculated by dividing the vote of the strongest losing candidate by his vote *plus* the vote of the weakest winning candidate. The opposition's percentage in the contest for the second seat was the following ratio: the total vote of all losing candidates, except the strongest, over this numerator plus the vote of the strongest winning candidate. In this way, contests in multimember districts were compared to those in single-member districts.

were based on the 1958–1964 elections—that is, the four preceding the period when major decisions affecting the poverty program were made.

Even this crude measure of organizational strength sharply differentiates the four cities, as shown in Table 1. Chicago's powerful organization so discouraged insurgent candidates that over three-fourths of the races were uncontested. The party's power to prevent contests declined somewhat in Philadelphia, fell even more in New York, and reached a low point in Los Angeles, where scarcely more than one-half the elections were uncontested. The impact of the regular organization's strength on the voters is further demonstrated in those elections which were contested. In only 8 per cent of all cases did opposition candidates in Chicago receive 20 per cent or more of the vote. This figure increased to 23 per cent in Philadelphia, to 33 per cent in New York and in Los Angeles to 43 per cent. Additional details in Table 1 emphasize these intercity differences still more, but it is clear that this quantitative index conforms to the qualitative data on the strength of the dominant party in the four cities.

We hypothesized that the rank order of cities on this index of power dispersion would be directly associated with the intercity variation in the distribution of power to the poor through community action. In order to measure this power distribution through community action we constructed an index based on the following three indicators: *first*, the percentage of neighborhood and minority group representatives on the city-

TABLE 1   PERCENTAGE OF VOTE CAST FOR LOSING CANDIDATES IN
DEMOCRATIC PRIMARIES FOR LOWER HOUSE OF STATE
LEGISLATURE (1958–1964)

Percentage Cast for Losing Candidates[34]	Possible Contests (per cent)			
	Chicago	Philadelphia	New York	Los Angeles
No opposition	77.5	67.1	61.2	51.7
1–19	14.5	9.9	6.1	5.2
20–39	8.0	20.4	16.5	20.7
40–59[a]	0.0	2.6	16.2	12.1
60–	0.0	0.0	0.0	10.3
Total	100.0	100.0	100.0	100.0
Number of possible contests	(138)	(152)	(260)	(58)

[a] In contests involving three or more candidates, the losers together can receive more than 50 per cent of the vote.

wide community action agencies; *second,* the relative influence of the city administration over the selection of these representatives; and *third,* the degree to which the representatives could be held accountable by an organized constituency.

Chicago officials rejected uncontrolled participation of representatives as unwarranted. By March, 1966, only 8 per cent (six of seventy-eight) of the members of the decision-making committee even formally represented the poor. Moreover, these representatives were chosen directly by the staff of the Chicago poverty program, providing only *controlled* representation. Even though OEO in the summer of 1966 insisted on twice this number of low-income representatives, no substantial change in the process of selection was implemented. The controlled selection process prevented the representatives from developing any autonomous relations with their constituency.

Direct elections were held in Philadelphia to select the representatives of the poor to twelve neighborhood committees. Each committee chose one representative to the city-wide committee, giving the poor 40 per cent representation on a thirty member body. The process of selection in Philadelphia was virtually *free of control* by either the city administration or the party organization. But it failed to encourage the *organization* of the poverty program's clients. Turnout in the first poverty election amounted to only 3 per cent of the eligible voters and the vote itself revealed a "friends-and-neighbors" pattern.[35] Without an organized constituency for whom the representatives could speak and to whom they could be held accountable, they lacked both the power and the continuing mandate to influence decisively either the poverty program or city politics in general. Developments in Los Angeles resembled the Philadelphia pattern. The poor were given 35 per cent representation on the policy-making body (seven out of twenty members). The representatives were chosen in an *uncontrolled* election but less than 1 per cent of the eligible electorate participated. The winners, once again, had only limited and *unorganized* contact with their constituents.

In New York under the Wagner administration, client representatives were entitled to thirty-two out of one hundred seats on the policy-making body and another ten seats were awarded to such neighborhood nonprofit organizations as Haryou-Act and Mobilization for Youth. But before the poor filled all these seats, which would have given them 42 per cent representation, the Lindsay administration in the summer of 1966 created

---

[35] The term is taken from V. O. Key, Jr., *Southern Politics* (New York: Alfred A. Knopf, 1949). Analysis of election data on Philadelphia's poverty election discloses the same tendency to vote for one's neighbor when organization is lacking that Key found in many southern states.

a new structure which gave 50 per cent representation to neighborhood organizations. Community conventions attended by delegates of neighborhood organizations were used to select the representatives of the poor; since the process was relatively free of official control, it encouraged *uncontrolled, yet organized* representation. While this process in some cases enabled established agencies and churches to influence the selection of representatives, more often it brought out leaders interested in developing new organizations with a protest orientation and militancy unrivalled in other cities. The leaders were at once free of official control yet capable to speak for a fairly well organized body of constituents.

Thus, Table 2 shows the New York program most fully realized the uncontrolled, but organized, representation of the poor. The greatest control over the representation occurred in Chicago, while Philadelphia and Los Angeles fell in between these two extremes.[36]

TABLE 2   PERCENTAGE AND TYPE OF REPRESENTATION OF
LOW-INCOME GROUPS ON COMMUNITY ACTION
AGENCIES' POLICY-MAKING COMMITTEES

City	Percentage Representation	Type of Representation
Chicago	8	City controlled
Los Angeles	35	Uncontrolled, but disorganized
Philadelphia	40	Uncontrolled, but disorganized
New York	50	Uncontrolled and organized

When we cross-tabulated the index of party strength against the propensity of community action to distribute power, the curvilinear relationship presented in Figure 1 emerged. Since we hypothesized a direct or linear relationship, this finding forced us to reconsider the intervening political processes in the four cities. As we shall show, this curvilinear relationship was produced by the unexpectedly complex effect of power dispersion on *the mayor's resources, the mayor's interests,* and *the flow of demand inputs.*

The stronger the party organization, the greater the resources of the mayor in bargaining with other political actors. The relations between the mayor and OEO provide a particularly graphic example. Chicago's

[36] Work now in progress on more precise indicators of the increase in the political power of the poor due to the community action program tentatively suggests that in fact there was more participation in Los Angeles than in Philadelphia. The less refined indicators of formal participation presented here were unable to distinguish between the two cities.

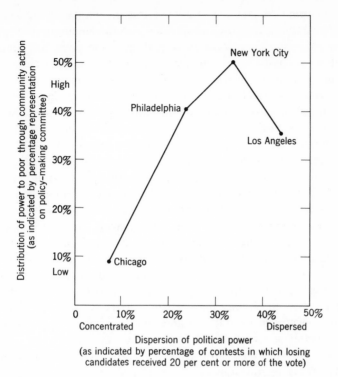

FIGURE 1   *Dispersion of political power and distribution of power to poor through community action.*

Mayor Daley, through his unified congressional delegation, could force OEO to withdraw suggestions that the poor be given meaningful representation. But in cities where power was more dispersed, the mayor had fewer resources in his negotiations with Washington. In New York and Los Angeles, where reform had been most successful, several congressmen publicly opposed the mayor's attempt to centralize power over the program. Thus, the mayor's political *resources* were a critical intervening variable. The relationship between power dispersion and the political *interests* of the mayor was more complicated. Where power was centralized, as in Chicago, the mayor's primary goal was to maintain the party organization which kept him in power. His machine, which had been the sole significant political force in low-income areas, did not welcome the growth of independent neighborhood organizations. Once such groups received federal poverty funds, they would have an interest in acquiring the political power to assure additional funding later. Thus politicized, they might begin to campaign for their friends and against their enemies, eventually entering their own candidates and threatening traditional party bastions. Saul Alinsky, a militant organizer of the poor, has argued that "City Hall obviously won't finance a group dedicated to fierce political independence and to the servicing of its own self-interests as it defines them."[37] Such considerations, while most prominent in Chicago, also affected the interests of Mayor Tate in Philadelphia and, to a lesser extent, Wagner in New York.

But with the election in New York of Republican-reform Mayor John Lindsay, the political interests of the city administration there were significantly altered. In fact, new community organizations in low-income neighborhoods are probably least threatening to reform mayors such as Lindsay, *who are elected over the opposition of a political machine that has not been completely destroyed.* Such mayors have depended largely on the newspapers, the prestigious civic associations, and middle-class reform clubs for support against the hostile party organizations. But they have lacked a stable organization to mobilize voters in low-income areas where a weakened but potentially troublesome party organization has existed and, as in St. Louis, may regain power. By supporting funds for community groups in low-income areas and listening to the demands they make for better services, reform mayors may well have been able to expand their own voting constituencies. At least they created new problems for hostile party organizations.

All of this seemed to apply with particular force if the reform mayor supporting participation of the poor was also a Republican, as was John Lindsay. The Democratic Negro or Spanish-speaking residents living in the lowest-income areas often demanded desegregated housing and education as well as more opportunities for employment as skilled workers. Such policies often threatened Democratic Italian, Polish, and Irish groups more than Republicans in silk-stocking neighborhoods.[38]

At least until 1966, the machine's distribution of material benefits to particular individuals enabled the Democratic leaders to ignore some of the Negroes' demands for improved public service, thus helping maintain the loyalty of white ethnic groups. But if new community organizations were to make more universalistic demands for better public services, it would not be as easy to maintain the Negroes' Democratic solidarity. Indeed, an alliance between Republicans and reformers, occurring in 1965 in both New York and Philadelphia, led to significant Negro defection

[37] Steven M. Loveday, *Wall Street Journal*, February 18, 1966.
[38] Because of this tension between low income whites and Negroes, Samuel Lubell argued in 1964 that "for some years to come the likely pattern of political conflict promises to be stormier at the local and state levels than at the presidential level." Samuel Lubell, *White and Black: Test of a Nation* (New York: Harper & Row, 1964), p. 160.

in local elections from the Democrats. In comparison with many of the white supporters of the Democratic regular party organizations, reform groups have an ideological commitment to the goals of integration and dispersal of political power to the poor. A further political link between the well-to-do and the poorest voting groups may lie in their "public-regarding" values toward major civic expenditures, as distinguished from the more "private-regarding" values held by Catholic ethnic groups.[39] Indeed, this coalition would parallel the tacit partnership between big business and Negro groups in many southern cities. In New York City the extensive but far from total dispersion of political power induced the mayor to encourage a program aimed at the political conditions of poverty.

In Los Angeles, unlike New York, power is so dispersed that the mayor has little to gain by a further distribution of power to low-income groups. The city's nonpartisan elections and utter organizational fluidity have so destroyed party structures as to eliminate them as political threats. No political considerations, in other words, offset the manifold ways in which autonomous community groups could create new administrative problems for the mayor of a large city.[40]

A curvilinear relationship also obtained between the dispersion of political power and the articulation of opposition demands. Centralized power in Chicago prevented opposition demands from flowing easily through the system. Newspapers and even politicians of opposite political persuasion did not give neighborhood groups their support, lest they antagonize the mayor needlessly. Even existing protest groups which thrive on public controversy had to concentrate their attention on one or two issues— education and housing—lest they overcommitted their limited resources. As Bachrach and Baratz have argued, this "other face of power" prevented controversial issues from arising.[41] In Philadelphia, and still more in New York, less concentrated political power enabled private welfare agencies, Republicans, reformers, and leading newspapers to support neighborhood groups seeking power for themselves. Encouraged by such support, the neighborhood groups formed city-wide *ad hoc* committees which gave organization and focus to the scattered demands of neighborhood groups. As the level of controversy rose, the mayors were forced to concede substantial representation to the poor, particularly in New York City.

In Los Angeles, on the other hand, the extreme dispersal of political power actually handicapped the demands of neighborhood groups for more power. The reform clubs had no local bosses to attack, which weakened

---

[39] James Q. Wilson and Edward C. Banfield, "Public-Regardingness as a Value Premise in Voting Behavior," *American Political Science Review,* LVIII (December, 1964), pp. 876–887.
[40] *Supra,* pp. 275–276.
[41] Peter Bachrach and Morton S. Baratz, "Two Faces of Power," *American Political Science Review,* LVI (December, 1962), pp. 947–952.

their interest in stimulating competing community organizations. Their candidate for mayor, James Roosevelt, gave significantly less support to representation of the poor than Ryan and Lindsay had in New York City. Similarly, with no machine left to reform, Los Angeles newspapers had little incentive to encourage participation. Consequently, they supported the incumbent mayor and opposed the neighborhood groups in the controversy. As a result, disorganization of the city's political and social structures left the opposition with few channels through which their demands could flow. More successful than their eastern counterparts, Los Angeles reformers so individualized power that, like Humpty Dumpty, it could not readily be put back together again—even to achieve reform purposes.

To summarize, in Chicago the confluence of three factors—the mayor's resources, his interests, and the flow of demands—all acted to reduce the distribution of power in the community action program. The mayor had both the interests and the resources to minimize representation for minority groups. Meanwhile, the opposition found it difficult to enlist political support for its demands. In New York City, power was distributed in the community action programs under Lindsay because the mayor's interests coincided with this policy. Even under Wagner the mayor's resources were so limited that demands for representation of the poor, which flowed easily through the political system, could not be ignored. Intermediate levels of power dispersion in the community action structures appeared in both Philadelphia and Los Angeles but for very different reasons. Although both mayors opposed extensive representation of the poor by autonomous organized groups, Mayor Tate in Philadelphia had far more resources than Mayor Yorty of Los Angeles with which to resist these demands. On the other hand, the availability of important allies provided protest groups in Philadelphia with channels for making their demands felt. In Los Angeles, these channels were unavailable. It is this *varying* combination of mayoral resources and interests together with the flow of political demands which produced the curvilinear relationship between existing power dispersion and the further distribution of power in the community action program.

INTERCITY VARIATION: THE ATTACK
ON ECONOMIC POVERTY

Our second hypothesis was that machine cities would obtain and expend community action dollars more quickly than reform cities. To test this hypothesis we again used party strength as our index of the dispersion

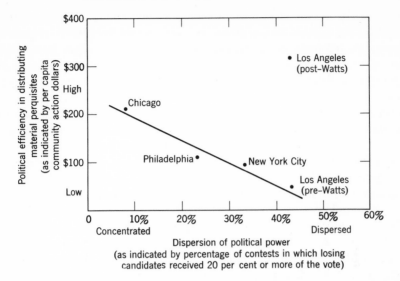

FIGURE 2   Dispersion of political power and political efficiency in distributing material perquisites.

of political power, cross-tabulating it this time against the community action dollars per poor family granted to the city by the OEO during the first two years of the program.[42] The relationship is presented in Figure 2. The per capita grant to each city at least provides a good measure of the speed with which each city initiated a relatively sizeable attack on poverty as an economic condition through OEO's program.[43]

Significantly enough, this quantitative measure correlates with the more qualitative and impressionistic data on the four local programs, particularly the speed and efficiency with which the community action activities were inaugurated. Thus Chicago, which received the highest amount per

[42] The amount of community action dollars granted to each city for the fiscal years 1964–1965 and 1965–1966 were obtained directly from the OEO. The number of poor families in each city is the number of families with incomes of less than $3000 per year given in the 1960 census.

[43] These figures by themselves do not directly support the contention that a particular city distributed more material services to the poor than any others. The city that received the most money per low income family may have spent a much higher percentage for the benefit of relatively advantaged rather than poor citizens. But since there was no persuasive evidence for this contention, it was more reasonable to assume that approximately the same per cent of poverty funds actually benefited the poor in all four cities. If this is so, then we may conclude that those cities that obtained more money from Washington did do more to ameliorate the *economic* needs of the poor.

poor family—$211—was able to put its neighborhood centers, the main focus of its program, into operation as early as March, 1965. Philadelphia, which secured only $112 in federal funds per low-income family, did not select an executive director until April, 1965, and the program did not begin until the following summer. Even then most money was channeled through other agencies, such as the schools and the recreation department, because the poverty agency was not able to establish its own program quickly. New York acted still more slowly since it received only $101 per poor family and used this amount only haphazardly and after long delays. Its first Progress Center was not opened until March, 1966—one year later than the Chicago center. At the end of the 1965–1966 fiscal year, $12 million remained undistributed. The most spectacular delays occurred in Los Angeles which, at the end of the first fiscal year, did even more poorly, having been granted less than half as much per poor family as any of the other three cities—only $25 per poor family (calculated at the two-year rate of $50 in Figure 2). During this period the community action program barely got under way in the face of the many-sided and complicated conflict that pervaded the first year of the program.

But this linear relationship between centralized political power and the distribution of material perquisites was dramatically altered by the social explosion in Los Angeles at the beginning of the second fiscal year of the program. During the second year the city became the most favored of the four, receiving $158 per poor family (calculated on the graph at the two year rate of $316). The Watts riots, the most violent of all Negro outbursts, so disturbed OEO officials that comparatively vast sums of money were allocated to the city. But in light of the poverty program's expressed goal of reducing urban violence, through the elimination of economic poverty, the disruption of the usual operation of structural variables, such as the dispersion of political power, was scarcely surprising.

But with this exception, the linear relationship did obtain between the centralization of political power in the city and the capacity of the authorities to secure and distribute material perquisites through the community action program. The intervening variables seem fairly clear. All mayors felt it was in their *interest* to obtain as much money as possible and to spend it as quickly as possible. In this way favorable publicity would replace carping criticism. The key variable thus became the resources of the mayor to attain this goal, which, as we have seen, depended directly on the dispersion of political power within the city. With centralized political power, the mayor could bargain effectively with OEO officials. In particular, Mayor Daley's cohesive congressional delegation placed him in an enviable position in comparison to his New York or Los Angeles counterpart. But of equal and perhaps greater importance, the mayor needed power to still internal dissent so that he could quickly establish this new program. One potential source of dissent was the demand for the representation of the poor on the board; where this controversy flourished, it delayed the implementation of the program. But other factors contributed to program delay as well. In Philadelphia, Mayor Tate delayed selecting his executive director because he was unable or unwilling to choose between competing Negro factions. In New York City, the opening of neighborhood centers and sites for Head Start, the educational program for preschool children, was delayed by other city bureaucracies too entrenched to be forced to cooperate by a politically weak mayor. The Los Angeles conflict combined tensions among competing governmental jurisdictions with the struggle between neighborhood organizations and the mayor.

In summary, the strength of party organization affected both the amount of federal funding and the speed and efficiency with which local agencies were established to disburse the funds. Since machines were historically dependent on the efficient payment of material benefits to their various supporters, it is not surprising to find that greater organizational strength increases the mayor's ability to obtain and distribute poverty funds. By contrast, the far greater concern of reform movements for "democracy" and "honesty" in political processes and structures, as opposed to the distribution of material outputs, explains the less effective distribution of funds by poverty agencies in cities with strong reform traditions.

DEMOCRACY IN URBAN AMERICA: POWER, PERQUISITES, AND "CLASS BIAS" RECONSIDERED

This investigation of the community action program sustains the points of agreement between the early critics of machine politics and its more recent scholarly defenders. The tendency of machine cities to use material perquisites to concentrate power (condemned by the early and heralded by the later scholarly tradition) directed the program toward solving the economic dimension of poverty. For the most part, the stronger the machine, the greater the tendency toward distributing material perquisites rather than power. The reform tradition's propensity to disperse power even at the cost of rapid implementation of governmental programs (endorsed by the early scholars but criticized by recent social scientists) led reform cities to concentrate their attack on the political conditions associated with poverty. For the most part, the stronger the reform movement, the easier it was for neighborhood groups to gain political influence but the harder for the city to inaugurate programs intended to alleviate the economic plight of the poor. It is worth emphasizing, however, that in Los Angeles—before the Watts riots—neither the political nor the economic conditions of poverty were significantly alleviated by the program. The complete triumph of reform seems to have reduced the political system's capacity to achieve even reformist goals.

If our empirical investigation validates the consensus of the two scholarly traditions on the different political processes of reform and machine politics, it remains to assess the class bias of these two structures with regard to the community action program. In fact, community action within the war on poverty provides a case where both reform and machine administrations were explicitly directing benefits to the poor. Their common pattern of upper-class bias, in other words, was measurably reduced. This relatively greater concern for the lower classes than in the past reflects the New Deal revolution in the American party system which made the dominant national party, the Democrats, responsive to the disadvantaged urban masses. Specifically, the Johnson administration provided that federal funds be explicitly directed *by law* to a low-income clientele. Moreover, the 90 per cent federal share of the cost spared substantial expense for local taxpayers, minimizing much potential community opposition. This trend has been reinforced by the growing numbers and sophistication of urban Negro and Spanish-speaking voters. Far more than earlier urban immigrants, these groups have begun to demand political solutions for their economic and social problems. But because of the present differences in the class biases of machine and reform politics, this change has not occurred to the same extent or in the same way in all four of our cities.

As an autonomous political organization determined to maintain its centralized power, pre-New Deal machines were reluctant to become too dependent on any one social stratum—even the politically powerful business class. They balanced upper-class money against lower-class votes. By contrast, the reformers, simply by dividing and distributing power among various social groups, produced a political process which more nearly reflected the general upper-class dominance of the social system. Unlike the lower classes in machine cities, those experiencing reform rule received less in the way of both psychic rewards (such as ethnic balance on electoral slates) and particularistic material benefits. But in this post-New Deal period, the machine's same concern to maintain its independence leads it to oppose an independent position of political power for minority and low-income groups despite the rise of the welfare state. As a result, low-income groups in reform cities may eventually use their greater power to secure more far-reaching collective material rewards, such as substantial increases in city services. Just as reform politics was *relatively* more favorable to upper classes in the past, it may now—at least potentially—be *relatively* more favorable than machine politics to low-income groups.

Admittedly, the significant participation of the poor requires an extreme effort on the part of poverty officials and neighborhood leaders. Even in New York City, participation of community groups was only beginning in 1966. Yet the city's program at times neared the point of complete administrative breakdown. At least in the short run, maximizing the value of participation appeared to come only at considerable cost to alleviating economic poverty. On the other hand, the long-run solutions to economic poverty may depend on increased political power of precisely these presently disorganized and politically unsophisticated low-income neighborhood groups. Rather than limited economic assistance over the short run, it may be more beneficial for the lower classes to stimulate neighborhood and minority groups to demand future economic assistance on a far more massive scale.

Finally, the varied experience of the community action program in the four cities shows that the conflict of basic political values between the two scholarly traditions over mass participation in political action parallels a continuing conflict within the American political system. As we have seen, this conflict produced a critical ambiguity in OEO's own goals. In Kornhauser's terms, the pessimistic adherents of contemporary pluralism within OEO sought to reduce the vulnerability of existing political and social elites to mass discontent and pressure by attacking the economic basis of poverty.[44] The more optimistic adherents of Jeffersonian participatory democracy sought to incorporate these same masses as effective actors in the political community by providing for the "maximum feasible participation" of the program's clientele.

We have tried to show that this ambivalence within OEO was substantially magnified by the differences imposed by the reform and machine traditions in the different cities. In a manner typical of federal programs in the decentralized American political system, OEO policies conformed to and thus reinforced the prevailing political pattern in each local community. Where the dominant party machine still had centralized political power, OEO concentrated on its goal of restraining and pacifying the lower class by improving its economic condition. In cities where an historically successful reform tradition had dispersed power, federal administrators partially realized their Jeffersonian goal of increasing citizen participation in political life.

The continuing impact of these historical differences on contemporary efforts to end urban poverty suggest how deeply rooted has been the

[44] William Kornhauser, *The Politics of Mass Society* (New York: The Free Press of Glencoe, 1959).

fear of the democratic masses in a large and necessarily impersonal republic. This presumably modern issue was not only raised by de Tocqueville in the age of Jackson, but it was also debated (although less explicitly) by the Hamiltonians and Jeffersonians of the preceding generation. Moreover, as our analysis points out, the issue of mass participation in political life was also a covert but perhaps the critical issue between machines and reformers in American urban politics even before it emerged as the mass society theory of many modern political philosophers in Europe and the United States. We have argued here that this theoretical question has reemerged as a significant political issue in contemporary American politics.

Index